Undergraduate Texts in Physics

Undergraduate Texts in Physics (UTP) publishes authoritative texts covering topics encountered in a physics undergraduate syllabus. Each title in the series is suitable as an adopted text for undergraduate courses, typically containing practice problems, worked examples, chapter summaries, and suggestions for further reading. UTP titles should provide an exceptionally clear and concise treatment of a subject at undergraduate level, usually based on a successful lecture course. Core and elective subjects are considered for inclusion in UTP.

UTP books will be ideal candidates for course adoption, providing lecturers with a firm basis for development of lecture series, and students with an essential reference for their studies and beyond.

Ray LaPierre

Getting Started in Quantum Optics

 Springer

Ray LaPierre
Hamilton, ON, Canada

ISSN 2510-411X　　　　　　　ISSN 2510-4128　(electronic)
Undergraduate Texts in Physics
ISBN 978-3-031-12434-1　　　　ISBN 978-3-031-12432-7　(eBook)
https://doi.org/10.1007/978-3-031-12432-7

This Springer imprint is published by the registered company Springer Nature Switzerland AG
The registered company address is: Gewerbestrasse 11, 6330 Cham, Switzerland

Preface

Getting Started in Quantum Optics was born from lecture notes developed for an introductory course in quantum optics offered to undergraduate students in the Department of Engineering Physics at McMaster University (Hamilton, Ontario, Canada). The book was written for students who have completed an introductory course on quantum mechanics and electromagnetism, but otherwise have no background in quantum optics. *Getting Started in Quantum Optics* is mostly intended as a self-contained introduction to the theory of quantum optics with the beginner in mind.

This book covers canonical quantization, the quantum harmonic oscillator, vacuum fluctuations, Fock states, the single photon state, the quantum optics treatment of the beam splitter and the interferometer, multimode quantized light, and coherent and incoherent states. The book provides a treatment of squeezed light and its use in the Laser Interferometer Gravitational-Wave Observatory (LIGO). The Heisenberg limit is described, along with NOON states and their application in super-sensitivity, super-resolution, and quantum lithography. Applications of entanglement and coincidence measurements are described including ghost imaging, quantum illumination, absolute photodetector calibration, and interaction-free measurement. Light-matter interaction, atomic clocks, and atom cooling and trapping are included. The book does not cover quantum computing, which is a topic treated by another book by the author [1]. Together with quantum computing, the topics of this book attempt to form an almost complete introductory description of the "second quantum revolution".

Since the book is intended for the undergraduate beginner, I try not to say "it is easy to show" too often. Rigorous derivations are given, although some calculations are completed as exercises in the book and heuristic arguments are occasionally employed to avoid getting bogged down too much in calculations. The emphasis is on physical understanding. Thus, in some instances, a simplification of the topics is presented for pedagogical reasons. For example, the canonical quantization of light is presented using entirely electric and magnetic fields, rather than the usual approach of using the vector potential. I do not use density matrices anywhere in this book. Students are expected to be familiar with the Dirac bra-ket notation and tensor product of states.

Reference

1. Ray LaPierre, *Introduction to Quantum Computing* (Springer, 2021). https://doi.org/10.1007/978-3-030-69318-3

Acknowledgments

I am grateful to McMaster University, the Department of Engineering Physics, my many colleagues, my students, and my family for inspiring me to write this book. Any errors in the book are entirely my own.

How to Use This Book

This book is intended for a single semester (~12 week) elective course on quantum optics, comprised of approximately 36 1-hour lectures (3 hours per week). Chapters are intended to be covered consecutively. Instructors may wish to begin with an overview of the bra-ket notation, inner product, expectation values, tensor product, and related topics. A suggested lecture schedule is:

Lecture 1: Chap. 1 – Canonical Quantization
Lecture 2–3: Chap. 2 – Quantum Harmonic Oscillator
Lecture 4–5: Chap. 3 – Canonical Quantization of Light
Lecture 6: Chap. 4 – Fock States and the Vacuum
Lecture 7: Chap. 5 – Single Photon State
Lecture 8–9: Chap. 6 – Single Photon on a Beam Splitter
Lecture 10–11: Chap. 7 – Single Photon in an Interferometer
Lecture 12: Chap. 8: Entanglement
Lecture 13–14: Chap. 9 – Multimode Quantized Radiation
Lecture 15–16: Chap. 10 – Coherent State
Lecture 17: Chap. 11 – Coherent State on a Beam Splitter
Lecture 18–19: Chap. 12 – Incoherent State
Lecture 20–21: Chap. 13 – Homodyne and Heterodyne Detection
Lecture 22–23: Chap. 14 – Coherent State in an Interferometer
Lecture 24–25: Chap. 15 – Squeezed Light
Lecture 26–27: Chap. 16 – Squeezed Light in an Interferometer
Lecture 28–29: Chap. 17 – Heisenberg Limit
Lecture 30–31: Chap. 18 – Quantum Imaging
Lecture 32–33: Chap. 19 – Light–Matter Interaction
Lecture 34: Chap. 20 – Atomic Clock
Lecture 35–36: Chap. 21 – Atom Cooling and Trapping

The book assumes that students have successfully completed an introductory course in quantum mechanics, which is typically in the second year of a 4-year undergraduate program in physics or related disciplines. Thus, this book is intended as a course for the third or fourth year of an undergraduate program, or the entry level of a graduate program.

Each chapter includes exercises which can be completed by the student as homework assignments or used for tutorial instruction. A solutions manual is available from the publisher for qualified instructors. Each chapter also includes references for more advanced study, and further reading is listed at the end of the book.

Contents

Chapter 1
Canonical Quantization

We begin with Hamiltonian mechanics and a method called "canonical quantization", developed by Paul Dirac, used for finding the Hamiltonian of a quantum system from its classical counterpart. Once you know the Hamiltonian of the quantum system, you can determine its quantum properties from the time-dependent Schrodinger equation. The canonical quantization procedure gives us the canonically conjugate variables of the system that satisfy a commutation relation, such as the Heisenberg uncertainty relation.

1.1 Hamiltonian Mechanics

Hamiltonian mechanics was formulated by William Rowan Hamilton in 1833. Hamiltonian mechanics is equivalent to Newton's laws of motion but provides a simplification of the analysis for many dynamical systems. Another approach is Lagrangian mechanics, which we leave to the reader as a topic for independent study. In Hamiltonian mechanics, a system is described by canonically conjugate variables denoted by q_i and p_i:

$$q_1, q_2, \ldots, q_i, \ldots; p_1, p_2, \ldots, p_i, \ldots \tag{1.1}$$

q_i and p_i are also called the generalized position and momentum coordinates, respectively. For example, q_1, q_2 and q_3 may refer to the actual position coordinates (x, y, z) of a particle and p_1, p_2 and p_3 correspond to its linear momentum (p_x, p_y and p_z). If there is more than one particle, then q_4, q_5, q_6, p_4, p_5 and p_6 are the corresponding variables for the second particle, and so on. In general, q_i and p_i may represent dynamic variables other than position and momentum, depending on the system. For example, to describe a pendulum (Exercise 1.1), it is easier to assign q_i as the angle of the pendulum and p_i as the angular momentum. The q_i and p_i variables, if they are canonically conjugate variables, satisfy the Hamilton equations:

© The Author(s), under exclusive license to Springer Nature Switzerland AG 2022
R. LaPierre, *Getting Started in Quantum Optics*, Undergraduate Texts in Physics,
https://doi.org/10.1007/978-3-031-12432-7_1

Fig. 1.1 A potential, $U(x)$

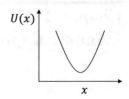

$$\frac{dq_i}{dt} = \frac{\partial H}{\partial p_i} \tag{1.2}$$

$$\frac{dp_i}{dt} = -\frac{\partial H}{\partial q_i} \tag{1.3}$$

where H is the Hamiltonian and t is the time. The Hamiltonian is the total energy of the system (kinetic energy plus potential energy) expressed in terms of the generalized coordinates.

To illustrate Hamilton's approach, let us find the equations of motion for a particle of mass m in a one-dimensional potential, $U(x)$, shown in Fig. 1.1. Although $U(x)$ is actually the potential energy, physicists often abbreviate this simply as "the potential". In this example, suppose the generalized coordinates (q_i, p_i) are the position (x) and momentum (p) of the particle:

$$q \to x \tag{1.4}$$

$$p \to m \frac{dx}{dt} \tag{1.5}$$

The Hamiltonian is the total energy (kinetic energy plus potential energy) expressed in terms of the generalized coordinates from Eqs. (1.4) and (1.5):

$$H = \frac{p^2}{2m} + U(x) \tag{1.6}$$

Using Eq. (1.6), the Hamilton equations become

$$\frac{dx}{dt} = \frac{\partial H}{\partial p} = \frac{p}{m} = v \tag{1.7}$$

$$\frac{dp}{dt} = -\frac{\partial H}{\partial x} = -\frac{\partial U}{\partial x} = F \tag{1.8}$$

where v is the velocity and F is the force. Equation (1.7) is simply the definition of momentum, while Eq. (1.8) reproduces the correct dynamical equation according to Newton's laws of motion. Thus, x and p satisfy the Hamilton equations, and we say that x and p are canonically conjugate variables.

Instead, suppose the generalized coordinates (q_i, p_i) are taken as the position (x) and velocity (v) of the particle:

$$q \rightarrow x \tag{1.9}$$

$$p \rightarrow v \tag{1.10}$$

The Hamiltonian expressed in terms of the generalized coordinates, Eqs. (1.9) and (1.10), is now

$$H = \frac{1}{2}mv^2 + U(x) \tag{1.11}$$

The Hamilton equations become

$$\frac{dx}{dt} = \frac{\partial H}{\partial v} = mv \text{ (incorrect)} \tag{1.12}$$

$$\frac{dv}{dt} = -\frac{\partial H}{\partial x} = -\frac{\partial U}{\partial x} = F \text{ (incorrect)} \tag{1.13}$$

Equations (1.12) and (1.13) are incorrect, since $\frac{dx}{dt} = v$ is clearly not equal to mv, and $\frac{dv}{dt}$ is not equal to F according to Newton's laws of motion. Thus, x and v are not canonically conjugate variables.

> **Exercise 1.1** Solve the equation of motion for a simple pendulum using Newton's laws. Repeat using Hamiltonian mechanics.

1.2 Canonical Quantization

Canonical quantization is a prescribed method of finding the Hamiltonian of a quantum system. The procedure was developed by Paul Dirac in 1925 (Fig. 1.2). Dirac proposed that any system for which we have a classical description can be quantized according to the procedure of canonical quantization. In canonical quantization, the generalized coordinates of the classical description, found by Hamilton's approach (described in the previous section), are replaced by the corresponding quantum operators (denoted by a "hat", $\widehat{}$):

$$H(q_1, \ldots, q_i, \ldots; p_1, \ldots, p_i, \ldots) \xrightarrow[\text{quantization}]{\text{canonical}} \widehat{H}(\widehat{q}_1, \ldots, \widehat{q}_i, \ldots; \widehat{p}_1, \ldots, \widehat{p}_i, \ldots) \tag{1.14}$$

Fig. 1.2 Paul Dirac (Nobel
Prize in Physics in 1933).
(Credit: Wikimedia
Commons [1])

where the classical description is on the left and the quantum description is on the right. The Hamiltonian of the quantum system, $\widehat{H}$, is expressed in terms of the generalized coordinates (now operators) on the right-hand side of Eq. (1.14). For example, according to Sect. 1.1, the generalized coordinates for a particle of mass m in a potential, $U(x)$, are x and p. The Hamiltonian for the corresponding quantum system becomes

$$H = \frac{p^2}{2m} + U(x) \xrightarrow[\quad\text{quantization}\quad]{\text{canonical}} \widehat{H} = \frac{\widehat{p}^2}{2m} + U(\widehat{x}) \qquad (1.15)$$

Once you know $\widehat{H}$ of the quantum system, you can determine its quantum properties from the time-dependent Schrodinger equation:

$$i\hbar \frac{\partial |\psi\rangle}{\partial t} = \widehat{H}|\psi\rangle \qquad (1.16)$$

where $|\psi\rangle$ is the state of the system and $\hbar$ is the reduced Planck constant ($\hbar = h/2\pi$). You may remember from introductory quantum mechanics that Eq. (1.16) reduces to the time-independent Schrodinger equation for stationary states:

$$\widehat{H}|\psi_n\rangle = E_n|\psi_n\rangle \qquad (1.17)$$

where E_n are the eigenenergies and $|\psi_n\rangle$ are the eigenstates (basis states) of the system.

1.3 Commutation Relations

Dirac showed that the canonically conjugate variables $(\hat{q}_i, \hat{p}_j)$ of the quantum system satisfy the commutation relation:

$$[\hat{q}_i, \hat{p}_j] = i\hbar\delta_{ij} \tag{1.18}$$

where δ_{ij} is the Kronecker function and, by definition,

$$[\hat{q}_i, \hat{p}_j] = \hat{q}_i\hat{p}_j - \hat{p}_j\hat{q}_i \tag{1.19}$$

Thus, when $i = j$, we say that $\hat{q}_i$ and $\hat{p}_i$ "do not commute"; that is, $[\hat{q}_i, \hat{p}_i] = \hat{q}_i\hat{p}_i - \hat{p}_i\hat{q}_i = i\hbar$. Otherwise, the operators commute. For example, when the generalized coordinates $(\hat{q}_i, \hat{p}_i)$ are the position and momentum, we have

$$[\hat{x}, \hat{p}_x] = i\hbar \tag{1.20}$$

$$[\hat{y}, \hat{p}_y] = i\hbar \tag{1.21}$$

$$[\hat{z}, \hat{p}_z] = i\hbar \tag{1.22}$$

Thus, position and momentum along the same direction do not commute (e.g., $\hat{x}$ and $\hat{p}_x$ do not commute), while position and momentum along different directions do commute (e.g., $\hat{x}$ and $\hat{p}_y$ commute). Eq. (1.20) leads to the well-known Heisenberg uncertainty relation:

$$\Delta x \Delta p_x \geq \frac{\hbar}{2} \tag{1.23}$$

with the same relation for the y and z directions arising from Eq. (1.21) and (1.22), respectively. In Eq. (1.23), Δx is the uncertainty in position x and Δp_x is the uncertainty in momentum along x. Uncertainty is defined as the standard deviation or root mean square (rms) error:

$$\Delta x = \sqrt{\langle (x - \langle x \rangle)^2 \rangle} \tag{1.24}$$

$$= \sqrt{\langle x^2 + \langle x \rangle^2 - 2x\langle x \rangle \rangle} \tag{1.25}$$

$$= \sqrt{\langle x^2 \rangle - \langle x \rangle^2} \tag{1.26}$$

where the brackets $\langle \rangle$ denote an average (in quantum mechanics, this is called the "expectation value" of x). In general, Eq. (1.26) gives the uncertainty in a measurable quantity x, and will be used frequently throughout this book. Note that $(\Delta x)^2$ is known as the variance of x.

Heisenberg's uncertainty relation (also known as the uncertainty principle) limits the accuracy with which a pair of canonically conjugate variables can be measured at the same time. Equation (1.23), for example, tells us that we cannot measure the position and momentum simultaneously to arbitrary accuracy. This has nothing to do with the accuracy of the instruments but is rather a fundamental limitation of nature. Conjugate variables that do not commute cannot be measured simultaneously without some minimum uncertainty. We will derive Eq. (1.23) in the next chapter in the case of the quantum harmonic oscillator, which is the most important quantum system in quantum optics.

Reference

1. https://commons.wikimedia.org/wiki/File:Paul_Dirac,_1933.jpg

Chapter 2
Quantum Harmonic Oscillator

The quantum harmonic oscillator (QHO) is introduced using the canonical quantization of the classical harmonic oscillator. An alternative formalism of the QHO due to Dirac is introduced along with the creation and annihilation operators. The expectation value and uncertainty of the position and momentum are derived, resulting in the Heisenberg uncertainty relation.

2.1 Classical Harmonic Oscillator

Consider a classical system comprised of a particle of mass, m, and position, x, moving in a one-dimensional parabolic potential:

$$U(x) = \frac{1}{2}kx^2 = \frac{1}{2}m\omega^2 x^2 \tag{2.1}$$

where k is a force constant and $\omega = \sqrt{k/m}$ is the angular frequency. The harmonic oscillator arises in a wide variety of classical systems, but most often as a mass on a spring described by Hooke's law ($F = -kx$). The generalized coordinates for this system are simply the position and momentum:

$$q \to x \tag{2.2}$$

$$p \to m\frac{dx}{dt} \tag{2.3}$$

and the Hamiltonian becomes

© The Author(s), under exclusive license to Springer Nature Switzerland AG 2022
R. LaPierre, *Getting Started in Quantum Optics*, Undergraduate Texts in Physics,
https://doi.org/10.1007/978-3-031-12432-7_2

$$H = \frac{p^2}{2m} + \frac{1}{2}m\omega^2 x^2 \tag{2.4}$$

The Hamilton equations become

$$\frac{dx}{dt} = \frac{\partial H}{\partial p} = \frac{p}{m} = v \tag{2.5}$$

$$\frac{dp}{dt} = -\frac{\partial H}{\partial x} = -m\omega^2 x = -\frac{\partial U}{\partial x} = F \tag{2.6}$$

The first equation is the definition of momentum ($p = mv$), while the second equation reproduces Newton's equation $\left(F = \frac{dp}{dt}\right)$. Thus, x and p satisfy the Hamilton equations (they give the correct dynamical behavior) and are therefore canonically conjugate variables.

Equations (2.5) and (2.6) are easily solved. Combining the two equations gives

$$\frac{d^2 x}{dt^2} = -\omega^2 x \tag{2.7}$$

with the solution

$$x = a\cos(\omega t + \varphi) \tag{2.8}$$

where the amplitude, a, and phase, φ, are determined by initial conditions. Equivalently, the solution may be written as

$$x = Ae^{-i\omega t} + c.c. \tag{2.9}$$

where $c.c.$ denotes the complex conjugate, and

$$A = \frac{a}{2}e^{-i\varphi} \tag{2.10}$$

Note that a is a real number, but A is a complex number. The negative sign in the exponent of Eq. (2.9) is by convention, although the positive exponent is also a valid solution.

2.2   Quantum Harmonic Oscillator

As described in Chap. 1, the Hamiltonian for the quantum harmonic oscillator (QHO) is obtained by canonical quantization where the coordinates (x, p) are replaced by their quantum operators:

$$\text{canonical}$$
$$H = \frac{p^2}{2m} + \frac{1}{2}m\omega^2 x^2 \xrightarrow{\quad\text{quantization}\quad} \hat{H} = \frac{\hat{p}^2}{2m} + \frac{1}{2}m\omega^2\hat{x}^2 \tag{2.11}$$

where $\hat{x}$ and $\hat{p}$ obey the commutation relation:

$$[\hat{x}, \hat{p}] = i\hbar \tag{2.12}$$

Equation (2.12) can be used to find the momentum operator $\hat{p}$ in terms of the x coordinate. Starting from Eq. (2.12) and according to the definition of the commutation relation:

$$(\hat{x}\hat{p} - \hat{p}\hat{x})|\psi\rangle = i\hbar|\psi\rangle \tag{2.13}$$

Expanding the left side of Eq. (2.13) gives

$$\hat{x}\hat{p}|\psi\rangle - \hat{p}\hat{x}|\psi\rangle = i\hbar|\psi\rangle \tag{2.14}$$

If $|\psi\rangle$ is in the position representation (i.e., $|\psi\rangle$ represents the familiar wavefunction, $\psi(x)$), then the operator $\hat{x}$ is simply the position, x; that is, $\hat{x}|\psi\rangle = x|\psi\rangle$. Thus, Eq. (2.14) becomes

$$x\hat{p}|\psi\rangle - \hat{p}x|\psi\rangle = i\hbar|\psi\rangle \tag{2.15}$$

In the second term on the left, $\hat{p}x|\psi\rangle$, we apply the rules of partial differentiation, that is, the operator $\hat{p}$ operates on x while keeping $|\psi\rangle$ constant, and then $\hat{p}$ operates on $|\psi\rangle$ while keeping x constant. This gives

$$x\hat{p}|\psi\rangle - (\hat{p}x)|\psi\rangle - x(\hat{p}|\psi\rangle) = i\hbar|\psi\rangle \tag{2.16}$$

In the second term on the left, $\hat{p}$ operates on x only. In the third term on the left, $\hat{p}$ operates on $|\psi\rangle$ only. The first and third terms on the left of Eq. (2.16) cancel, giving

$$-(\hat{p}x)|\psi\rangle = i\hbar|\psi\rangle \tag{2.17}$$

or

$$\hat{p}x = -i\hbar \tag{2.18}$$

The solution to Eq. (2.18) is

$$\hat{p} = \frac{\hbar}{i}\frac{\partial}{\partial x} \tag{2.19}$$

Equation (2.19) is easily checked by substitution in Eq. (2.18). Thus, the left-hand side of Eq. (2.18) is obtained by applying the differential operator $\hat{p}$ to x, giving $\frac{\hbar}{i}\frac{\partial}{\partial x}(x) = \frac{\hbar}{i}$, which is identical to the right-hand side of Eq. (2.18). Eq. (2.19) is a relationship that should be familiar from introductory quantum mechanics—it is the momentum in the position representation.

Using Eq. (2.19), the Hamiltonian in Eq. (2.11) can be written entirely in terms of the position, x:

$$\hat{H} = -\frac{\hbar^2}{2m}\frac{d^2}{dx^2} + \frac{1}{2}m\omega^2 x^2 \tag{2.20}$$

Finally, the stationary eigenstates and eigenenergies of the QHO can be found by substituting the Hamiltonian of Eq. (2.20) into the time-independent Schrodinger equation, Eq. (1.17), giving

$$-\frac{\hbar^2}{2m}\frac{d^2}{dx^2}|\psi_n\rangle + \frac{1}{2}m\omega^2 x^2|\psi_n\rangle = E_n|\psi_n\rangle \tag{2.21}$$

The solution to Eq. (2.21) can be found in any introductory textbook on quantum mechanics. The eigenstates are

$$|\psi_n\rangle = \frac{1}{\sqrt{2^n n!}}\left(\frac{m\omega}{\pi\hbar}\right)^{\frac{1}{4}} e^{-\frac{1}{2}\xi^2} H_n(\xi) \tag{2.22}$$

where $H_n(\xi)$ are Hermite polynomials of degree n; $n = 0, 1, 2, \ldots$; and $\xi = \left(\frac{m\omega}{\hbar}\right)^{1/2}x$. The first few Hermite polynomials, $H_n(\xi)$, are

$$H_0 = 1 \tag{2.23}$$

$$H_1 = 2\xi \tag{2.24}$$

$$H_2 = 4\xi^2 - 2 \tag{2.25}$$

$$H_3 = 8\xi^3 - 12\xi \tag{2.26}$$

$$H_4 = 16\xi^4 - 48\xi^2 + 12 \tag{2.27}$$

$$H_5 = 32\xi^5 - 160\xi^3 + 120\xi \tag{2.28}$$

The eigenenergies resulting from Eq. (2.21) are

$$E_n = \left(n + \frac{1}{2}\right)\hbar\omega\ldots, \; n = 0,1,2,\ldots \tag{2.29}$$

Fig. 2.1 The first few
eigenenergies and
eigenstates (wavefunctions)
of the quantum harmonic
oscillator (QHO)

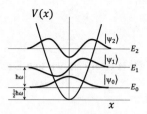

The first few eigenstates and eigenenergies of the QHO are shown in Fig. 2.1. An
important feature of the QHO is the existence of a minimum energy (ground state
energy), $E_0 = \frac{1}{2} \hbar\omega$, corresponding to $n = 0$ in Eq. (2.29).

2.3 Dirac Formalism

Paul Dirac formulated an alternative approach to solve the QHO. Suppose $\widehat{H}$ can be
factorized as follows:

$$\widehat{H} = \widehat{O}^\dagger \widehat{O} + E_0 \tag{2.30}$$

where $\widehat{O}$ is some operator and $\widehat{O}^\dagger$ is the Hermitian conjugate. If $|\psi_n\rangle$ is an eigenstate
of $\widehat{H}$, then the eigenenergies are

$$E_n = \langle \psi_n | \widehat{H} | \psi_n \rangle \tag{2.31}$$

Substituting Eq. (2.30) for $\widehat{H}$ gives

$$\begin{aligned} E_n &= \langle \psi_n | (\widehat{O}^\dagger \widehat{O} + E_0) | \psi_n \rangle \\ &= \langle \psi_n | \widehat{O}^\dagger \widehat{O} | \psi_n \rangle + E_0 \end{aligned} \tag{2.32}$$

This means:

$$E_n \geq E_0 \tag{2.33}$$

If $\widehat{O} | \psi_0 \rangle = 0$, then the minimum energy (ground state energy, E_0) is found.
At this point, it is helpful to define dimensionless operators, $\widehat{Q}$ and $\widehat{P}$:

$$\widehat{Q} = \sqrt{\frac{m\omega}{\hbar}} \, \widehat{x} \tag{2.34}$$

$$\widehat{P} = \sqrt{\frac{1}{m\hbar\omega}}\,\widehat{p} \tag{2.35}$$

It is easily shown that Eqs. (2.11) and (2.12) become

$$\widehat{H} = \frac{\hbar\omega}{2}\left(\widehat{Q}^2 + \widehat{P}^2\right) \tag{2.36}$$

$$\left[\widehat{Q}, \widehat{P}\right] = i \tag{2.37}$$

To perform the factorization of $\widehat{H}$, like Eq. (2.30), we can rewrite $\widehat{H}$ in terms of $\widehat{Q}$ and $\widehat{P}$:

$$\widehat{H} = \hbar\omega\left[\frac{1}{\sqrt{2}}\left(\widehat{Q} - i\widehat{P}\right)\frac{1}{\sqrt{2}}\left(\widehat{Q} + i\widehat{P}\right) + \frac{1}{2}\right] \tag{2.38}$$

Note that, according to Eq. (2.37), $\widehat{Q}$ and $\widehat{P}$ do not commute, which leads to the factor of ½ in Eq. (2.38).

Exercise 2.1 Derive Eq. (2.38).

Equation (2.38) inspires us to define two new operators:

$$\widehat{a} = \frac{1}{\sqrt{2}}\left(\widehat{Q} + i\widehat{P}\right) \tag{2.39}$$

$$\widehat{a}^\dagger = \frac{1}{\sqrt{2}}\left(\widehat{Q} - i\widehat{P}\right) \tag{2.40}$$

where $\widehat{a}$ is called the creation or raising operator and $\widehat{a}^\dagger$ is called the annihilation, destruction or lowering operator. The reason for these names will become clear later. With these definitions, the Hamiltonian in Eq. (2.38) becomes

$$\widehat{H} = \hbar\omega\left(\widehat{a}^\dagger\widehat{a} + \frac{1}{2}\right) \tag{2.41}$$

We have succeeded in factoring $\widehat{H}$, similar to Eq. (2.30). We will see below that $\widehat{a}|\psi_0\rangle = 0$, so Eq. (2.41) gives the ground state energy of $E_0 = \frac{1}{2}\hbar\omega$, identical to that from Eq. (2.29).

Note that Eq. (2.41) does not contain $\widehat{x}$ or $\widehat{p}$. One of the beautiful aspects of Dirac's formalism is that the expression for $\widehat{H}$ in Eq. (2.41) is applicable to any type of oscillator—for example, the quantization of atomic vibrations in a crystal (phonons), the quantization of flux and charge in an LC circuit oscillator, and the quantization of the electromagnetic field (photons). The last example is the topic of the next chapter.

Using $\widehat{Q}$ and $\widehat{P}$, we can also derive the commutation relation for $\widehat{a}$ and $\widehat{a}^\dagger$:

$$
\begin{aligned}
\left[\widehat{a},\ \widehat{a}^\dagger\right] &= \widehat{a}\,\widehat{a}^\dagger - \widehat{a}^\dagger\widehat{a} \\
&= \left[\frac{1}{\sqrt{2}}\left(\widehat{Q} + i\widehat{P}\right)\frac{1}{\sqrt{2}}\left(\widehat{Q} - i\widehat{P}\right) - \frac{1}{\sqrt{2}}\left(\widehat{Q} - i\widehat{P}\right)\frac{1}{\sqrt{2}}\left(\widehat{Q} + i\widehat{P}\right)\right] \\
&= \frac{1}{2}\left[\left(\widehat{Q}^2 - i\widehat{Q}\,\widehat{P} + i\widehat{P}\,\widehat{Q} + \widehat{P}^2\right) - \left(\widehat{Q}^2 + i\widehat{Q}\,\widehat{P} - i\widehat{P}\,\widehat{Q} + \widehat{P}^2\right)\right] \\
&= -i\left(\widehat{Q}\,\widehat{P} - \widehat{P}\,\widehat{Q}\right) \\
&= -i\left[\widehat{Q},\ \widehat{P}\right] \\
&= -i(i) \\
&= 1
\end{aligned}
$$

where we have used Eq. (2.37). Thus, we arrive at the important result:

$$
\left[\widehat{a}, \widehat{a}^\dagger\right] = 1 \tag{2.42}
$$

or, by definition:

$$
\widehat{a}\,\widehat{a}^\dagger - \widehat{a}^\dagger\widehat{a} = 1 \tag{2.43}
$$

Thus, $\widehat{a}$ and $\widehat{a}^\dagger$ do not commute. Eq. (2.43) will have far-reaching consequences throughout this book.

Note that, by definition, Eqs. (2.39) and (2.40) indicate that $\widehat{a} \neq \widehat{a}^\dagger$. Thus, $\widehat{a}$ and $\widehat{a}^\dagger$ are not Hermitian. This means that $\widehat{a}$ and $\widehat{a}^\dagger$ do not correspond to any physical observable (Exercise 2.2). However, we can form observables from combinations of $\widehat{a}$ and $\widehat{a}^\dagger$. For example, $\widehat{Q}$ and $\widehat{P}$ can be derived in terms of $\widehat{a}$ and $\widehat{a}^\dagger$. By rearranging Eqs. (2.39) and (2.40), we obtain

$$
\widehat{Q} = \frac{1}{\sqrt{2}}\left(\widehat{a} + \widehat{a}^\dagger\right) \tag{2.44}
$$

$$
\widehat{P} = \frac{-i}{\sqrt{2}}\left(\widehat{a} - \widehat{a}^\dagger\right) \tag{2.45}
$$

$\widehat{Q} = \widehat{Q}^\dagger$ and $\widehat{P} = \widehat{P}^\dagger$, so $\widehat{Q}$ and $\widehat{P}$ are Hermitian and may correspond to observables (position and momentum, respectively).

Exercise 2.2 Show that operators corresponding to physical observables must be Hermitian.

2.4   Number Operator

The number operator is defined as

$$\widehat{N} = \widehat{a}^\dagger \widehat{a} \tag{2.46}$$

Thus, we can express the Hamiltonian for the QHO as

$$\widehat{H} = \hbar\omega\left(\widehat{a}^\dagger \widehat{a} + \frac{1}{2}\right) = \hbar\omega\left(\widehat{N} + \frac{1}{2}\right) \tag{2.47}$$

For simplicity and by convention, let us adopt a new notation for the eigenstate of a QHO, replacing $|\psi_n\rangle$ with the ket, $|n\rangle$. $|n\rangle$ is called the number state or Fock state, named after the Russian theorist, Vladimir Fock. Thus,

$$\widehat{H}|n\rangle = E_n|n\rangle \tag{2.48}$$

Substituting Eq. (2.47) for $\widehat{H}$ and Eq. (2.29) for E_n, we obtain

$$\hbar\omega\left(\widehat{N} + \frac{1}{2}\right)|n\rangle = \hbar\omega\left(n + \frac{1}{2}\right)|n\rangle \tag{2.49}$$

Thus, canceling $\frac{1}{2}\hbar\omega|n\rangle$ on both sides of Eq. (2.49), we get

$$\widehat{N}|n\rangle = n|n\rangle \tag{2.50}$$

The number operator, $\widehat{N} = \widehat{a}^\dagger \widehat{a}$, operating on $|n\rangle$ gives the eigenvalue, n. Any eigenstate of $\widehat{N}$ with eigenvalue n is also an eigenstate of $\widehat{H}$ with eigenvalue $E_n = \hbar\omega\left(n + \frac{1}{2}\right)$. The states $|n\rangle$ form an orthonormal basis:

$$\langle n|m\rangle = \delta_{nm} \tag{2.51}$$

where n and $m = 0, 1, 2, \ldots$. Also, the states $|n\rangle$ form a complete basis, meaning any state $|\psi\rangle$ of the QHO can be written as a superposition of the $|n\rangle$ states.

As an illustration of the Dirac formalism, let us find the ground state wavefunction for the QHO. Applying Eq. (2.50) with $n = 0$, we get

$$\widehat{N}|0\rangle = 0$$

or

$$\widehat{a}^\dagger \widehat{a}|0\rangle = 0 \tag{2.52}$$

Applying the bra, $\langle 0|$, to both sides gives the inner product:

$$\langle 0 | \hat{a}^\dagger \hat{a} | 0 \rangle = 0 \tag{2.53}$$

or, equivalently, the modulus is zero:

$$\| \hat{a} | 0 \rangle \| = 0 \tag{2.54}$$

Since the eigenstates are normalized ($\| | 0 \rangle \| = 1$), we must have

$$\hat{a} | 0 \rangle = 0 \tag{2.55}$$

Substituting $\hat{a} = \frac{1}{\sqrt{2}} \left(\hat{Q} + i \hat{P} \right)$, $\hat{Q} = \sqrt{\frac{m\omega}{\hbar}} \hat{x}$, $\hat{P} = \sqrt{\frac{1}{mh\omega}} \hat{p}$, $\hat{x} = x$, and $\hat{p} = \frac{\hbar}{i} \frac{\partial}{\partial x}$ into Eq. (2.55) gives the differential equation:

$$\left(x + \frac{\hbar}{m\omega} \frac{\partial}{\partial x} \right) | 0 \rangle = 0 \tag{2.56}$$

The solution to Eq. (2.56) is

$$| 0 \rangle = \left(\frac{m\omega}{\pi\hbar} \right)^{\frac{1}{4}} e^{-\frac{1}{2} \frac{m\omega}{\hbar} x^2} \tag{2.57}$$

which can be verified by substitution into Eq. (2.56). Equation (2.57) is the correct ground state wavefunction of the QHO, according to Eq. (2.22) with $n = 0$. Also, as mentioned previously, $\hat{a} | 0 \rangle = 0$, which gives the ground state energy of $E_0 = \frac{1}{2} \hbar\omega$ from Eq. (2.47).

The Dirac formalism presents an alternative approach to solve the QHO as compared to Sect. 2.2. At the end of Sect. 2.7, we will see how to generate the excited state wavefunctions (i.e., $| 1 \rangle$, $| 2 \rangle$, ...). First, we need some more formalism.

2.5   Annihilation Operator

In the previous section, we found $\hat{a} | 0 \rangle = 0$. What does the general case, $\hat{a} | n \rangle$, give? Starting with the definition of the number operator, $\hat{N} | n \rangle = n | n \rangle$, we have

$$\hat{a}^\dagger \hat{a} | n \rangle = n | n \rangle \tag{2.58}$$

Applying $\hat{a}$ to both sides gives

$$\hat{a} \hat{a}^\dagger \hat{a} | n \rangle = \hat{a} n | n \rangle \tag{2.59}$$

Using the commutation relation, Eq. (2.43) gives

$$(\hat{a}^\dagger \hat{a} + 1)\hat{a}|n\rangle = \hat{a}n|n\rangle \tag{2.60}$$

Rearranging, we get

$$(\hat{a}^\dagger \hat{a})\hat{a}|n\rangle = (n - 1)\,\hat{a}|n\rangle \tag{2.61}$$

or

$$\hat{N}\,\hat{a}|n\rangle = (n - 1)\hat{a}|n\rangle \tag{2.62}$$

Thus, $\hat{a}|n\rangle$ is an eigenstate of $\hat{N}$ with eigenvalue $n - 1$. In other words, $\hat{a}|n\rangle$ gives the state $|n - 1\rangle$:

$$\hat{a}|n\rangle \rightarrow |n - 1\rangle \tag{2.63}$$

For this reason, $\hat{a}$ is called the lowering or annihilation operator. It lowers n to $n - 1$. In other words, it destroys one quantum of excitation of the QHO.

Is $\hat{a}|n\rangle$ normalized? Let us suppose that $\hat{a}|n\rangle$ produces the state $|n - 1\rangle$ with coefficient (normalization factor) c_n:

$$|\psi\rangle = \hat{a}|n\rangle = c_n|n - 1\rangle \tag{2.64}$$

Using $|\psi\rangle = \hat{a}|n\rangle$, the inner product $\langle\psi|\psi\rangle$ is

$$\langle\psi|\psi\rangle = \langle n|\hat{a}^\dagger \hat{a}|n\rangle = n \tag{2.65}$$

Alternatively, using $|\psi\rangle = c_n|n - 1\rangle$, the inner product $\langle\psi|\psi\rangle$ is

$$\langle\psi|\psi\rangle = \langle n - 1|c_n^* c_n|n - 1\rangle = |c_n|^2 \tag{2.66}$$

where c_n^* is the complex conjugate of c_n. Equating the two inner products in Eqs. (2.65) and (2.66) gives the normalization factor:

$$c_n = \sqrt{n} \tag{2.67}$$

Thus, from Eq. (2.64)

$$\hat{a}|n\rangle = \sqrt{n}|n - 1\rangle \tag{2.68}$$

2.6   Creation Operator

In the previous section, we found that the annihilation operator, $\hat{a}|n\rangle$, gives the state $|n-1\rangle$ with normalization factor $\sqrt{n}$. What about $\hat{a}^\dagger|n\rangle$? Again, let us start with the definition of the number operator, $\hat{N}|n\rangle = n|n\rangle$, or

$$\hat{a}^\dagger\hat{a}\,|n\rangle = n|n\rangle \tag{2.69}$$

Applying $\hat{a}^\dagger$ to both sides gives

$$\hat{a}^\dagger\hat{a}^\dagger\hat{a}\,|n\rangle = \hat{a}^\dagger n|n\rangle \tag{2.70}$$

Using the commutation relation, Eq. (2.43) gives

$$\hat{a}^\dagger(\hat{a}\,\hat{a}^\dagger - 1)|n\rangle = \hat{a}^\dagger n|n\rangle \tag{2.71}$$

Rearranging, we get

$$\hat{a}^\dagger\hat{a}\,\hat{a}^\dagger|n\rangle = (n+1)\hat{a}^\dagger|n\rangle \tag{2.72}$$

or

$$\hat{N}\,\hat{a}^\dagger|n\rangle = (n+1)\hat{a}^\dagger|n\rangle \tag{2.73}$$

Thus, $\hat{a}^\dagger|n\rangle$ is an eigenstate of $\hat{N}$ with eigenvalue $n+1$. In other words, $\hat{a}^\dagger|n\rangle$ gives the state $|n+1\rangle$:

$$\hat{a}^\dagger|n\rangle \rightarrow |n+1\rangle \tag{2.74}$$

$\hat{a}^\dagger$ is called the raising or creation operator. It raises n to $n+1$. In other words, $\hat{a}^\dagger$ creates one quantum of excitation of the QHO.

Is $\hat{a}^\dagger|n\rangle$ normalized? Let us suppose that $\hat{a}^\dagger|n\rangle$ produces the state $|n+1\rangle$ with coefficient (normalization factor) c_n:

$$|\psi\rangle = \hat{a}^\dagger|n\rangle = c_n|n+1\rangle \tag{2.75}$$

Using $|\psi\rangle = \hat{a}^\dagger|n\rangle$, the inner product $\langle\psi|\,\psi\rangle$ is

$$\langle\psi|\psi\rangle = \langle n|\hat{a}\hat{a}^\dagger|n\rangle = \langle n|(\hat{a}^\dagger\hat{a} + 1)|n\rangle = n+1 \tag{2.76}$$

Alternatively, using $|\psi\rangle = c_n|n+1\rangle$, the inner product $\langle\psi|\psi\rangle$ is

$$\langle \psi | \psi \rangle = \langle n+1 | c_n^* c_n | n+1 \rangle = |c_n|^2 \tag{2.77}$$

Equating the two inner products, Eqs. (2.76) and (2.77), gives the normalization factor:

$$c_n = \sqrt{n+1} \tag{2.78}$$

Thus, from Eq. (2.75),

$$\hat{a}^\dagger | n \rangle = \sqrt{n+1} \, | n+1 \rangle \tag{2.79}$$

2.7 Creating Excited States from the Ground State

Starting with the $|0\rangle$ state, we can create any excited state using multiple applications of the creation operator:

$$\widehat{N}(\hat{a}^\dagger)^n |0\rangle = n \, (\hat{a}^\dagger)^n |0\rangle \tag{2.80}$$

Equation (2.80) states that $(\hat{a}^\dagger)^n |0\rangle$ is an eigenstate of the number operator, $\widehat{N}$, with eigenvalue, n. Also, $(\hat{a}^\dagger)^n |0\rangle$ is an eigenstate of $\widehat{H}$ with eigenvalue $E_n = \hbar\omega \left(n + \frac{1}{2}\right)$.

Let us prove Eq. (2.80). To do so, we need to arrange all $\hat{a}^\dagger$ operators to the left of all $\hat{a}$ operators. This arrangement is called the normal order. This is a commonly used trick in quantum optics that will be found throughout this book (indeed, we already used this trick in Eqs. (2.60) and (2.76)). The normal ordering is done by repeated application of the commutation relation, $\hat{a}\hat{a}^\dagger = \hat{a}^\dagger\hat{a} + 1$. We start with the left side of Eq. (2.80) and apply the definition of $\widehat{N} = \hat{a}^\dagger\hat{a}$:

$$\widehat{N}(\hat{a}^\dagger)^n |0\rangle = \left(\hat{a}^\dagger\hat{a}\right)\underbrace{\hat{a}^\dagger\hat{a}^\dagger\cdots\hat{a}^\dagger}_{n \text{ times}}|0\rangle \tag{2.81}$$

$$= \hat{a}^\dagger\left(\hat{a}\,\hat{a}^\dagger\right)\underbrace{\hat{a}^\dagger\hat{a}^\dagger\cdots\hat{a}^\dagger}_{n-1 \text{ times}}|0\rangle \tag{2.82}$$

In Eq. (2.81), "n times" means that $\hat{a}^\dagger$ is repeated n times, and similarly for subsequent equations. Using the commutation relation, $\hat{a}\,\hat{a}^\dagger = \hat{a}^\dagger\hat{a} + 1$, gives

$$\widehat{N}\left(\widehat{a}^\dagger\right)^n|0\rangle = \widehat{a}^\dagger\left(\widehat{a}^\dagger\widehat{a}+1\right)\underbrace{\widehat{a}^\dagger\widehat{a}^\dagger\cdots\widehat{a}^\dagger}_{n-1\ \text{times}}|0\rangle \tag{2.83}$$

$$= \widehat{a}^\dagger\widehat{a}^\dagger\widehat{a}\underbrace{\widehat{a}^\dagger\widehat{a}^\dagger\cdots\widehat{a}^\dagger}_{n-1\ \text{times}}|0\rangle + \underbrace{\widehat{a}^\dagger\widehat{a}^\dagger\cdots\widehat{a}^\dagger}_{n\ \text{times}}|0\rangle \tag{2.84}$$

If we repeat the normal ordering on the first term of Eq. (2.84), we get

$$\widehat{N}\left(\widehat{a}^\dagger\right)^n|0\rangle = \widehat{a}^\dagger\widehat{a}^\dagger\left(\widehat{a}^\dagger\widehat{a}+1\right)\underbrace{\widehat{a}^\dagger\widehat{a}^\dagger\cdots\widehat{a}^\dagger}_{n-2\ \text{times}}|0\rangle + \underbrace{\widehat{a}^\dagger\widehat{a}^\dagger\cdots\widehat{a}^\dagger}_{n\ \text{times}}|0\rangle \tag{2.85}$$

$$= \widehat{a}^\dagger\widehat{a}^\dagger\widehat{a}^\dagger\widehat{a}\underbrace{\widehat{a}^\dagger\widehat{a}^\dagger\cdots\widehat{a}^\dagger}_{n-2\ \text{times}}|0\rangle + \underbrace{\widehat{a}^\dagger\widehat{a}^\dagger\cdots\widehat{a}^\dagger}_{n\ \text{times}}|0\rangle + \underbrace{\widehat{a}^\dagger\widehat{a}^\dagger\cdots\widehat{a}^\dagger}_{n\ \text{times}}|0\rangle \tag{2.86}$$

We see that the last two terms in Eq. (2.86) are identical. Each application of the normal ordering shifts the annihilation operator $\widehat{a}$ once toward the right and adds the term $\underbrace{\widehat{a}^\dagger\widehat{a}^\dagger\cdots\widehat{a}^\dagger}_{n\ \text{times}}|0\rangle$. Repeated application of the normal ordering procedure eventually results in

$$\widehat{N}(\widehat{a}^\dagger)^n|0\rangle = \widehat{a}^\dagger\widehat{a}^\dagger\cdots\widehat{a}|0\rangle + n\underbrace{\widehat{a}^\dagger\widehat{a}^\dagger\cdots\widehat{a}^\dagger}_{n\ \text{times}}|0\rangle \tag{2.87}$$

Finally, the first term becomes 0 (since $\widehat{a}|0\rangle = 0$), giving

$$\widehat{N}(\widehat{a}^\dagger)^n|0\rangle = n\underbrace{\widehat{a}^\dagger\widehat{a}^\dagger\cdots\widehat{a}^\dagger}_{n\ \text{times}}|0\rangle \tag{2.88}$$

$$= n(\widehat{a}^\dagger)^n|0\rangle \tag{2.89}$$

Thus, we have proven Eq. (2.80).

Is $|\psi\rangle = (\widehat{a}^\dagger)^n|0\rangle$ normalized? That is, does $\langle\psi|\psi\rangle = 1$?:

$$\langle\psi|\psi\rangle = \langle 0|\underbrace{\widehat{a}\,\widehat{a}\cdots\widehat{a}}_{n\ \text{times}}\underbrace{\widehat{a}^\dagger\widehat{a}^\dagger\cdots\widehat{a}^\dagger}_{n\ \text{times}}|0\rangle \tag{2.90}$$

$$= \langle 0|\underbrace{\widehat{a}\,\widehat{a}\cdots\widehat{a}}_{n-1\ \text{times}}\left(\widehat{a}\widehat{a}^\dagger\right)\underbrace{\widehat{a}^\dagger\widehat{a}^\dagger\cdots\widehat{a}^\dagger}_{n-1\ \text{times}}|0\rangle \tag{2.91}$$

Applying the commutation relation to the middle $\widehat{a}\,\widehat{a}^\dagger$ pair:

$$\langle \psi | \psi \rangle = \langle 0 | \underbrace{\widehat{a}\,\widehat{a}\cdots\widehat{a}}_{n-1 \text{ times}} \left(\widehat{a}^\dagger \widehat{a} + 1 \right) \underbrace{\widehat{a}^\dagger \widehat{a}^\dagger \cdots \widehat{a}^\dagger}_{n-1 \text{ times}} | 0 \rangle \tag{2.92}$$

$$= \langle 0 | \underbrace{\widehat{a}\,\widehat{a}\cdots\widehat{a}}_{n-1 \text{ times}} \left(\widehat{a}^\dagger \widehat{a} \right) \underbrace{\widehat{a}^\dagger \widehat{a}^\dagger \cdots \widehat{a}^\dagger}_{n-1 \text{ times}} | 0 \rangle + \langle 0 | \underbrace{\widehat{a}\,\widehat{a}\cdots\widehat{a}\widehat{a}}_{n-1 \text{ times}} \underbrace{\widehat{a}^\dagger \widehat{a}^\dagger \cdots \widehat{a}^\dagger}_{n-1 \text{ times}} | 0 \rangle \tag{2.93}$$

Repeated application (n times) of the normal ordering procedure results in

$$\langle \psi | \psi \rangle = n \, \langle 0 | \underbrace{\widehat{a}\,\widehat{a}\cdots\widehat{a}}_{n-1 \text{ times}} \underbrace{\widehat{a}^\dagger \widehat{a}^\dagger \cdots \widehat{a}^\dagger}_{n-1 \text{ times}} | 0 \rangle \tag{2.94}$$

Now we repeat the normal ordering procedure again, resulting in

$$\langle \psi | \psi \rangle = n(n-1) \, \langle 0 | \underbrace{\widehat{a}\,\widehat{a}\cdots\widehat{a}}_{n-2 \text{ times}} \underbrace{\widehat{a}^\dagger \widehat{a}^\dagger \cdots \widehat{a}^\dagger}_{n-2 \text{ times}} | 0 \rangle \tag{2.95}$$

Continuing in this manner, we eventually get

$$\langle \psi | \psi \rangle = n(n-1)(n-2)\ldots(1) = n! \tag{2.96}$$

where $n!$ refers to the factorial of n. Hence, the normalized state is

$$| n \rangle = \frac{\left(\widehat{a}^\dagger \right)^n}{\sqrt{n!}} | 0 \rangle \tag{2.97}$$

Exercise 2.3 Starting from Eq. (2.97), show that $\widehat{a} | n \rangle = \sqrt{n} \, | n - 1 \rangle$, identical to Eq. (2.68).

Exercise 2.4 Starting from Eq. (2.97), show that $\widehat{a}^\dagger | n \rangle = \sqrt{n+1} \, | n + 1 \rangle$, identical to Eq. (2.79).

Starting with Eq. (2.57) for the wavefunction of the $| 0 \rangle$ state, we can generate the wavefunction for the excited states of the QHO using Eq. (2.97):

$$| n \rangle = \frac{1}{\sqrt{n!}} \left(\widehat{a}^\dagger \right)^n \left(\frac{m\omega}{\pi\hbar} \right)^{\frac{1}{4}} e^{-\frac{1}{2} \frac{m\omega}{\hbar} x^2} \tag{2.98}$$

Substituting $\widehat{a}^\dagger = \frac{1}{\sqrt{2}} \left(\widehat{Q} - i\widehat{P} \right)$, $\widehat{Q} = \sqrt{\frac{m\omega}{\hbar}} \, \widehat{x}$, $\widehat{P} = \sqrt{\frac{1}{m\hbar\omega}} \, \widehat{p}$, $\widehat{x} = x$, and $\widehat{p} = \frac{\hbar}{i} \frac{\partial}{\partial x}$ into Eq. (2.98) gives the differential equation:

$$|n\rangle = \frac{1}{\sqrt{2^n n!}} \left(\frac{m\omega}{\pi\hbar}\right)^{\frac{1}{4}} \left[\left(\sqrt{\frac{m\omega}{\hbar}} x - \sqrt{\frac{\hbar}{m\omega}} \frac{\partial}{\partial x}\right)\right]^n e^{-\frac{1}{2}\frac{m\omega}{\hbar} x^2} \tag{2.99}$$

If we introduce our prior notation of $\xi = \left(\frac{m\omega}{\hbar}\right)^{\frac{1}{2}} x$, we get

$$|n\rangle = \frac{1}{\sqrt{2^n n!}} \left(\frac{m\omega}{\pi\hbar}\right)^{\frac{1}{4}} \left[\left(\xi - \frac{\partial}{\partial \xi}\right)\right]^n e^{-\frac{1}{2}\xi^2} \tag{2.100}$$

By induction, you can show that

$$\left(\xi - \frac{\partial}{\partial \xi}\right)^n e^{-\frac{1}{2}\xi^2} = (-1)^n e^{\frac{1}{2}\xi^2} \frac{\partial^n}{\partial \xi^n} e^{-\xi^2} \tag{2.101}$$

Using Eq. (2.101), we can write Eq. (2.100) as

$$|n\rangle = \frac{1}{\sqrt{2^n n!}} \left(\frac{m\omega}{\pi\hbar}\right)^{\frac{1}{4}} (-1)^n e^{\frac{1}{2}\xi^2} \frac{\partial^n}{\partial \xi^n} e^{-\xi^2} \tag{2.102}$$

$$= \frac{1}{\sqrt{2^n n!}} \left(\frac{m\omega}{\pi\hbar}\right)^{\frac{1}{4}} e^{-\frac{1}{2}\xi^2} H_n(\xi) \tag{2.103}$$

where

$$H_n(\xi) = (-1)^n e^{\xi^2} \frac{\partial^n}{\partial \xi^n} e^{-\xi^2} \tag{2.104}$$

You can show that Eq. (2.104) generates Eqs. (2.23), (2.24), (2.25), (2.26), (2.27) and (2.28). Eq. (2.103) is identical to Eq. (2.22). Thus, we have generated the excited states for the QHO using the Dirac formalism.

2.8   Expectation Value and Uncertainty

The average or expectation value of Q (dimensionless position) for the QHO in the ground state, $|0\rangle$, is

$$\langle Q_0 \rangle = \langle 0|\hat{Q}|0\rangle = \frac{1}{\sqrt{2}} \langle 0|(\hat{a} + \hat{a}^\dagger)|0\rangle = 0 \tag{2.105}$$

since $\hat{a}|0\rangle = 0$ and $\langle 0|\hat{a}^\dagger = (\hat{a}|0\rangle)^\dagger = 0$. $\langle Q_0 \rangle = 0$ makes sense because the average position of a harmonic oscillator is indeed zero.

Although the average position is zero, the mean squared position of a harmonic oscillator is not zero. The average of Q^2 for the QHO in the ground state is

$$\langle Q_0{}^2 \rangle = \langle 0|\hat{Q}^2|0 \rangle = \frac{1}{2} \langle 0|(\hat{a}\,\hat{a} + \hat{a}\,\hat{a}^\dagger + \hat{a}^\dagger\hat{a} + \hat{a}^\dagger\hat{a}^\dagger)|0 \rangle \qquad (2.106)$$

The first term $\langle 0|\hat{a}\,\hat{a}|0 \rangle = 0$ since $\hat{a}|0 \rangle = 0$. The third term $\langle 0|\hat{a}^\dagger\hat{a}|0 \rangle = \langle 0|\hat{N}|0 \rangle = 0$. The fourth term $\langle 0|\hat{a}^\dagger\hat{a}^\dagger|0 \rangle = 0$, since the creation operators generate the state $|2 \rangle$, and $|0 \rangle$ and $|2 \rangle$ are orthogonal ($\langle 0|\,2 \rangle = 0$). This leaves only one of the four terms in Eq. (2.106), giving

$$\langle Q_0{}^2 \rangle = \frac{1}{2} \langle 0|\hat{a}\,\hat{a}^\dagger|0 \rangle \qquad (2.107)$$

Using the commutation relation for the normal ordering gives

$$\langle Q_0{}^2 \rangle = \frac{1}{2} \langle 0|(\hat{a}^\dagger\hat{a} + 1)|0 \rangle = \frac{1}{2} \qquad (2.108)$$

since $\langle 0|\hat{a}^\dagger\hat{a}|0 \rangle = \langle 0|\hat{N}|0 \rangle = 0$.

Using Eqs. (2.105) and (2.108), the uncertainty of Q for the QHO in the ground state is (recall Eq. (1.26))

$$\Delta Q_0 = \sqrt{\langle Q_0{}^2 \rangle - \langle Q_0 \rangle^2} = \sqrt{\frac{1}{2}} \qquad (2.109)$$

The square of an uncertainty, such as $(\Delta Q_0)^2$, is called the variance. Eq. (2.109) is also called the standard deviation or root-mean-square deviation. It gives the spread or dispersion in the possible values of Q_0.

Similarly, the average or expectation value of P (dimensionless momentum) for the QHO in the ground state is

$$\langle P_0 \rangle = 0 \qquad (2.110)$$

Again, this makes sense because the average momentum of a harmonic oscillator is indeed zero. Although the average momentum is zero, the mean squared momentum of a harmonic oscillator is not zero. The average of P^2 for the QHO in the ground state is

$$\langle P_0{}^2 \rangle = \frac{1}{2} \qquad (2.111)$$

Finally, the uncertainty in P for the QHO in the ground state is

$$\Delta P_0 = \sqrt{\langle P_0{}^2 \rangle - \langle P_0 \rangle^2} = \sqrt{\frac{1}{2}} \tag{2.112}$$

Exercise 2.5 Derive Eqs. (2.110) and (2.111).

2.9 Heisenberg Uncertainty Relation

Using Eqs. (2.109) and (2.112), we find

$$\Delta Q_0 \Delta P_0 = \frac{1}{2} \tag{2.113}$$

or, using Eqs. (2.34) and (2.35)

$$\Delta x_0 \Delta p_0 = \frac{\hbar}{2} \tag{2.114}$$

In general, for the state $|n\rangle$

$$\Delta Q = \sqrt{\langle Q_n{}^2 \rangle - \langle Q_n \rangle^2} = \sqrt{n + \frac{1}{2}} \tag{2.115}$$

$$\Delta P = \sqrt{\langle P_n{}^2 \rangle - \langle P_n \rangle^2} = \sqrt{n + \frac{1}{2}} \tag{2.116}$$

and

$$\Delta Q \Delta P = n + \frac{1}{2} \tag{2.117}$$

Equation (2.117) indicates that

$$\Delta Q \Delta P \geq \frac{1}{2} \tag{2.118}$$

with the equality for the ground state ($n = 0$). Hence, the Gaussian wavefunction (Eq. (2.57)), corresponding to the ground state ($n = 0$), is the state of the QHO with minimum uncertainty.

Exercise 2.6 Derive Eqs. (2.115) and (2.116).

Using Eqs. (2.34) and (2.35), we obtain

$$\Delta x \Delta p \geq \frac{\hbar}{2} \tag{2.119}$$

Equation (2.119) is the Heisenberg uncertainty relation, familiar from introductory quantum mechanics. Note that the ground state of the QHO ($n = 0$) has the smallest uncertainty permitted by the Heisenberg uncertainty relation. The uncertainty relation is consistent with the existence of a zero-point energy E_0. We can never have a harmonic oscillator with zero energy, because that would require its position and momentum to both have the precise value of zero (according to Eq. (2.11)), in contradiction to Eq. (2.119).

2.10 Some Important Relations

Below, we summarize the important results of this chapter for the QHO that will be used frequently throughout this book:

$$\text{Number operator}: \widehat{N} = \widehat{a}^\dagger \widehat{a} \tag{2.120}$$

$$\widehat{N}|n\rangle = n|n\rangle, n = 0,1,2,\ldots \tag{2.121}$$

$$\text{Hamiltonian}, \widehat{H} = \hbar\omega\left(\widehat{a}^\dagger \widehat{a} + \frac{1}{2}\right) \tag{2.122}$$

$$\widehat{H}|n\rangle = E_n|n\rangle \tag{2.123}$$

$$E_n = \hbar\omega\left(n + \frac{1}{2}\right) \tag{2.124}$$

$$\text{Commutation relation}: \left[\widehat{a}, \widehat{a}^\dagger\right] = 1 \tag{2.125}$$

$$|n\rangle = \frac{\left(\widehat{a}^\dagger\right)^n}{\sqrt{n!}}|0\rangle \tag{2.126}$$

$$\widehat{a}|n\rangle = \sqrt{n}|n-1\rangle \tag{2.127}$$

$$\widehat{a}^\dagger|n\rangle = \sqrt{n+1}\,|n+1\rangle \tag{2.128}$$

$$\widehat{a}|0\rangle = 0 \tag{2.129}$$

$$\text{Orthonormality}: \langle n|m\rangle = \delta_{nm} \tag{2.130}$$

Further information can be obtained from many introductory quantum mechanics books [1, 2] or the Further Reading provided at the end of this book.

References

1. B.H. Bransden and C.J. Joachain, *Introduction to quantum mechanics* (John Wiley & Sons, 1989).
2. H.C. Ohanian, *Principles of quantum mechanics* (Prentice Hall, 1990).

Chapter 3
Canonical Quantization of Light

Albert Einstein (Fig. 3.1) proposed the existence of light quanta in a series of publications beginning in 1905 [1–5], which he famously used to explain the photoelectric effect. The "first quantization" was the quantum mechanics of particles developed in the 1920s by Heisenberg, Schrodinger, Dirac, and others. The "second quantization" refers to the canonical quantization of the electromagnetic field, that is, the quantization of light. The second quantization is generally presented using the vector potential, $\widehat{A}$. We leave this to the reader as a topic for independent study. Instead, we adopt a simpler analysis here using only the electric and magnetic fields. In this chapter, the canonical quantization of light is presented by analogy to a quantum harmonic oscillator. The electric field operator is derived along with the quadrature operators. The standard quantum limit is introduced.

3.1 Single Mode of Radiation

Maxwell's equations provide a classical description of electric and magnetic fields. Maxwell's equations in free space (charge density $\rho = 0$, current density $J = 0$) are

$$\boldsymbol{\nabla} \cdot \boldsymbol{E} = 0 \tag{3.1}$$

$$\boldsymbol{\nabla} \cdot \boldsymbol{B} = 0 \tag{3.2}$$

$$\boldsymbol{\nabla} \times \boldsymbol{E} = -\frac{\partial \boldsymbol{B}}{\partial t} \tag{3.3}$$

$$\boldsymbol{\nabla} \times \boldsymbol{B} = \frac{1}{c^2} \frac{\partial \boldsymbol{E}}{\partial t} \tag{3.4}$$

where c is the speed of light in vacuum.

© The Author(s), under exclusive license to Springer Nature Switzerland AG 2022
R. LaPierre, *Getting Started in Quantum Optics*, Undergraduate Texts in Physics,
https://doi.org/10.1007/978-3-031-12432-7_3

Fig. 3.1 Albert Einstein
(Nobel Prize in Physics in
1921). (Credit: Wikimedia
Commons [6])

Fig. 3.2 Expression for the
classical electric field and
corresponding coordinate
system for a single mode

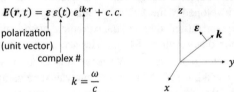

An oscillating electromagnetic field $(E(r, t), B(r, t))$ with angular frequency ω that satisfies Maxwell's equations is called a mode. The simplest possible mode is a travelling plane monochromatic electromagnetic wave in vacuum. The latter is called a "single mode" of radiation, which refers to an electromagnetic field of a single angular frequency, ω, and polarization, $\boldsymbol{\varepsilon}$. The classical electric field of a single mode can be expressed as

$$E(r, t) = \boldsymbol{\varepsilon}\, \varepsilon(t)\, e^{ik \cdot r} + c.c. \tag{3.5}$$

where $E(r, t)$ is the field vector, $\boldsymbol{\varepsilon}$ is a unit vector describing the polarization of the electric field, $\varepsilon(t)$ is a complex number describing the field amplitude that includes the time-dependence and initial phase of the electric field, k is the wavevector ($|k| = k = \omega/c$), and c.c. is the complex conjugate that makes $E(r, t)$ real. Here, $\boldsymbol{\varepsilon}$ is a real vector for simplicity, describing linear polarization. In general, $\boldsymbol{\varepsilon}$ is a complex number, needed to describe circular or elliptical polarization. Figure 3.2 illustrates the single mode field and the corresponding coordinate system.

Similarly, the magnetic field of a single mode is given by

$$B(r, t) = \frac{k \times \varepsilon}{\omega} \, \varepsilon(t) \, e^{ik \cdot r} + c.c. \tag{3.6}$$

E, B and k are mutually orthogonal (transverse waves), and E and B are in phase.

Exercise 3.1 Check that Eqs. (3.5) and (3.6) satisfy Maxwell's equations if the time-dependence is given by $\varepsilon(t) = e^{-i\omega t}$.

For simplicity, consider a single mode of radiation propagating in the x direction with polarization $\varepsilon = y$. Equations (3.5) and (3.6) simplify to

$$E(r, t) = y \, \varepsilon(t) \, e^{ikx} + c.c. \tag{3.7}$$

$$B(r, t) = \frac{1}{c} z \, \varepsilon(t) \, e^{ikx} + c.c. \tag{3.8}$$

Here, y and z are unit vectors in the y and z directions, respectively. By definition, $\nabla \times B$ is given by

$$\nabla \times B = \left[\left(\frac{\partial B_z}{\partial y} - \frac{\partial B_y}{\partial z} \right) x + \left(\frac{\partial B_x}{\partial z} - \frac{\partial B_z}{\partial x} \right) y + \left(\frac{\partial B_y}{\partial x} - \frac{\partial B_x}{\partial y} \right) z \right] \tag{3.9}$$

Substituting Eq. (3.8) gives

$$\nabla \times B = (0 - 0)x + \left\{ 0 - \frac{ik}{c} \left[\varepsilon(t) e^{ikx} + c.c. \right] \right\} y + (0 - 0)z \tag{3.10}$$

or, using $k = \omega/c$:

$$\nabla \times B = -i \left(\frac{\omega}{c^2} \right) E \tag{3.11}$$

Comparing Eq. (3.11) with Eq. (3.4) gives

$$\frac{\partial E}{\partial t} = -i\omega E \tag{3.12}$$

or, since the time-dependence is contained entirely in $\varepsilon(t)$, Eq. (3.12) gives

$$\frac{d\varepsilon(t)}{dt} = -i\omega\varepsilon(t) \tag{3.13}$$

with the solution:

$$\varepsilon(t) = \varepsilon^1 e^{-i(\omega t + \varphi)} \tag{3.14}$$

where φ is a phase determined by the initial conditions, and ε^1 is the field amplitude. The superscript "1" reminds us that the field is a single mode (single frequency, ω). Note that the positive exponential, $\varepsilon(t) = \varepsilon^1 e^{+i(\omega t + \varphi)}$, is also a solution to Eq. (3.13), although Eq. (3.14) is usually adopted by convention and describes a wave travelling in the $+x$ direction in Eq. (3.7).

3.2   Quadrature Components

Rather than writing $\varepsilon(t)$ with an amplitude and phase, as in Eq. (3.14), we may write it as a complex number with a real and imaginary component:

$$\varepsilon(t) = i\varepsilon^1 \frac{1}{\sqrt{2}}(Q + iP) \tag{3.15}$$

where Q and P are dimensionless real numbers. The i and $\frac{1}{\sqrt{2}}$ in Eq. (3.15) are present by convention. ε^1 is a constant with units of electric field. Q and P are real numbers and are dynamical variables describing the time-dependence of the field. Equation (3.15) is summarized in Fig. 3.3.

Substituting the complex field amplitude, Eq. (3.15), into Eq. (3.5) gives

$$
\begin{aligned}
\boldsymbol{E}(\boldsymbol{r}) &= i\varepsilon\varepsilon^1 \left(\frac{1}{\sqrt{2}}(Q + iP)e^{i\boldsymbol{k}\cdot\boldsymbol{r}} - \frac{1}{\sqrt{2}}(Q - iP)e^{-i\boldsymbol{k}\cdot\boldsymbol{r}} \right) \\
&= \varepsilon\sqrt{2}\varepsilon^1 \left(-Q\frac{(e^{i\boldsymbol{k}\cdot\boldsymbol{r}} - e^{-i\boldsymbol{k}\cdot\boldsymbol{r}})}{2i} - P\frac{(e^{i\boldsymbol{k}\cdot\boldsymbol{r}} + e^{-i\boldsymbol{k}\cdot\boldsymbol{r}})}{2} \right) \\
&= -\varepsilon\sqrt{2}\varepsilon^1 [Q \sin(\boldsymbol{k}\cdot\boldsymbol{r}) + P \cos(\boldsymbol{k}\cdot\boldsymbol{r})]
\end{aligned}
\tag{3.16}
$$

where we have used the Euler relation. Alternatively, we may write

$$\boldsymbol{E}(\boldsymbol{r}) = -\varepsilon\sqrt{2}\varepsilon^1 \sqrt{Q^2 + P^2} \sin(\boldsymbol{k} \cdot \boldsymbol{r} + \varphi) \tag{3.17}$$

where $\varphi = \tan^{-1}(P/Q)$. Hence, according to Eq. (3.16), Q and P are two components of the field that are out of phase by $\pi/2$. Q and P are known as the quadrature

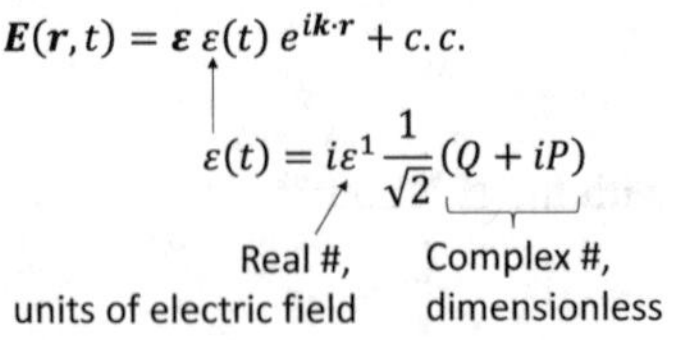

Fig. 3.3 Expression for the complex amplitude of a classical single mode field

$$\boldsymbol{E}(\boldsymbol{r},t) = \boldsymbol{\varepsilon}\, \varepsilon(t)\, e^{i\boldsymbol{k}\cdot\boldsymbol{r}} + c.c.$$

$$\varepsilon(t) = i\varepsilon^1 \frac{1}{\sqrt{2}}(Q + iP)$$

Real #,   Complex #,
units of electric field   dimensionless

components of the field. The field associated with a single mode of radiation may be described by two independent components—its magnitude and phase in Eq. (3.5) or, alternatively, by its two quadrature components (Q and P) in Eq. (3.16).

Exercise 3.2 Show Eqs. (3.16) and (3.17) are equivalent.

3.3 Classical Hamiltonian

The classical expression for the energy of an electromagnetic field is

$$H = \frac{1}{2} \int_V \left(\epsilon_o |\boldsymbol{E}|^2 + \frac{1}{\mu_o} |\boldsymbol{B}|^2 \right) dV \qquad (3.18)$$

where ϵ_o is the permittivity of free space, μ_o is the permeability of free space, and V is the volume of integration. ϵ_o and μ_o satisfy the relation $c = (\epsilon_o \mu_o)^{-1/2}$ where c is the speed of light in vacuum. Here, $\frac{\epsilon_o}{2} |\boldsymbol{E}|^2$ and $\frac{1}{2\mu_o} |\boldsymbol{B}|^2$ is the energy density (energy per unit volume) of the electric and magnetic fields, respectively. Using Eqs. (3.5) and (3.6), the electric field and magnetic field contributions to the energy are equal, giving

$$H = \epsilon_o \int_V |\boldsymbol{E}|^2 dV \qquad (3.19)$$

which is a well-known result for classical electromagnetic waves. Equation (3.17) gives

$$H = 2\epsilon_o \left(\epsilon^1 \right)^2 (Q^2 + P^2) \int_V \sin^2(\boldsymbol{k} \cdot \boldsymbol{r} + \varphi) dV \qquad (3.20)$$

We need to choose a volume for the integration of Eq. (3.20). The volume V could be a real volume (light confined in a cavity by two mirrors), the finite volume associated with a wavepacket or pulse of light, or waves confined in a fictitious box with periodic boundary conditions. The latter is a commonly used trick in quantum optics (indeed, throughout physics) and is discussed below.

To evaluate the integral in Eq. (3.20), we imagine the light confined in a fictitious box with side length L. The traveling wave satisfies periodic boundary conditions for the field, $\boldsymbol{E}(x = 0) = \boldsymbol{E}(x = L)$, as shown for a few modes in Fig. 3.4 (similarly for the y and z directions). We imagine the field approaching the right side of the box and "wrapping around" to the left side of the box. We can approach a continuous range of wavelengths by choosing a very large box (large L). With these boundary

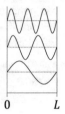

Fig. 3.4 Periodic boundary conditions for the field along the x, y, or z direction, leading to quantized values of the wavevector $k_x = \frac{2\pi}{L} n_x$, $k_y = \frac{2\pi}{L} n_y$ and $k_z = \frac{2\pi}{L} n_z$ where $n_{x,\,y,\,z}$ is an integer

conditions, the integration of $\sin^2(\mathbf{k} \cdot \mathbf{r} + \varphi)$ over the volume of the box becomes $\frac{1}{2}$, giving

$$H = \epsilon_o V \left(\varepsilon^1\right)^2 \left(Q^2 + P^2\right) \tag{3.21}$$

where $V = L^3$ is the volume of the box.

By analogy to Eq. (2.36), we can think of a single mode field as a harmonic oscillator where we found

$$H = \frac{\hbar\omega}{2} \left(Q^2 + P^2\right) \tag{3.22}$$

Equating Eqs. (3.21) and (3.22) gives

$$\varepsilon^1 = \sqrt{\frac{\hbar\omega}{2\epsilon_o V}} \tag{3.23}$$

ε^1 is called the "one photon amplitude of the mode with frequency ω". The concept of the photon will be introduced later.

3.4   Canonical Quantization

Now that we have the classical Hamiltonian, Eq. (3.22), we can perform the canonical quantization presented in Chap. 1. Let us define the canonical variables,

$$q = \sqrt{\hbar}\, Q \tag{3.24}$$

$$p = \sqrt{\hbar}\, P \tag{3.25}$$

The Hamiltonian in Eq. (3.22) becomes

$$H = \frac{\omega}{2} \left(q^2 + p^2\right) \tag{3.26}$$

The Hamilton equations give

$$\frac{dq}{dt} = \frac{\partial H}{\partial p} = \omega p \tag{3.27}$$

$$\frac{dp}{dt} = -\frac{\partial H}{\partial q} = -\omega q \tag{3.28}$$

Equations (3.27) and (3.28) are two coupled differential equations describing the dynamics for light. Do they give the expected dynamics described by Eq. (3.13)? Let us combine the two equations into a single differential equation. Multiplying Eq. (3.28) by i and adding Eq. (3.27) gives

$$\frac{d}{dt}(q + ip) = -i\omega(q + ip) \tag{3.29}$$

or, from Eqs. (3.15), (3.24) and (3.25):

$$\frac{d}{dt}\varepsilon(t) = -i\omega\,\varepsilon(t) \tag{3.30}$$

which is identical to Eq. (3.13). Thus, q and p give the correct dynamical equation. According to the canonical quantization procedure, $\hat{q}$ and $\hat{p}$ are quantum operators and are canonically conjugate variables. As canonically conjugate variables, $\hat{q}$ and $\hat{p}$ obey the commutation relation:

$$[\hat{q}, \hat{p}] = i\hbar \tag{3.31}$$

and, from Eq. (3.26), we can write the quantum Hamiltonian as

$$\hat{H} = \frac{\omega}{2}\left(\hat{q}^2 + \hat{p}^2\right) \tag{3.32}$$

Equivalently, using Eqs. (3.24) and (3.25), we can express $\hat{H}$ in terms of the dimensionless operators, $\hat{Q}$ and $\hat{P}$:

$$\hat{H} = \frac{\hbar\omega}{2}\left(\hat{Q}^2 + \hat{P}^2\right) \tag{3.33}$$

and

$$\left[\hat{Q}, \hat{P}\right] = i \tag{3.34}$$

Here, the operators $\hat{Q}$ and $\hat{P}$ are analogous to Eqs. (2.34) and (2.35) for the quantum harmonic oscillator (QHO). However, $\hat{Q}$ and $\hat{P}$ do not correspond to position and momentum of a photon! Here, $\hat{Q}$ and $\hat{P}$ correspond to two components of the electric field that are $\pi/2$ out of phase (quadrature components of the field), analogous to position and momentum $(\hat{x}, \hat{p})$ in a mechanical oscillator.

At this point, we can recall the earlier definition of $\hat{a}$ and $\hat{a}^\dagger$ and the commutation relation from Chap. 2:

$$\hat{a} = \frac{1}{\sqrt{2}}\left(\hat{Q} + i\hat{P}\right) \tag{3.35}$$

$$\hat{a}^\dagger = \frac{1}{\sqrt{2}}\left(\hat{Q} - i\hat{P}\right) \tag{3.36}$$

$$\left[\hat{a}, \hat{a}^\dagger\right] = 1 \tag{3.37}$$

Using Eqs. (3.35), (3.36) and (3.37) in Eq. (3.33), we obtain

$$\hat{H} = \hbar\omega\left(\hat{a}^\dagger\hat{a} + \frac{1}{2}\right) \tag{3.38}$$

which is the familiar Hamiltonian for the QHO. A more rigorous derivation of Eq. (3.38) using the vector potential and the normal modes of the field in a finite volume gives the same result (for more information on the latter approach, refer to the Further Reading provided at the end of this book).

Recalling the classical field, we have

$$\boldsymbol{E}(\boldsymbol{r}, t) = \boldsymbol{\varepsilon}\,\varepsilon(t)\,e^{i\boldsymbol{k}\cdot\boldsymbol{r}} + c.c. \tag{3.39}$$

with

$$\varepsilon(t) = i\varepsilon^1 \frac{1}{\sqrt{2}}(Q + iP) \tag{3.40}$$

Thus,

$$\boldsymbol{E}(\boldsymbol{r}, t) = i\boldsymbol{\varepsilon}\varepsilon^1\left(\frac{1}{\sqrt{2}}(Q + iP)e^{i\boldsymbol{k}\cdot\boldsymbol{r}} - \frac{1}{\sqrt{2}}(Q - iP)e^{-i\boldsymbol{k}\cdot\boldsymbol{r}}\right) \tag{3.41}$$

According to the canonical quantization, the quantum field is obtained by replacing the classical variables, Q and P, with the quadrature operators, $\hat{Q}$ and $\hat{P}$:

$$\hat{\boldsymbol{E}}(\boldsymbol{r}) = i\boldsymbol{\varepsilon}\varepsilon^1\left(\frac{1}{\sqrt{2}}\left(\hat{Q} + i\hat{P}\right)e^{i\boldsymbol{k}\cdot\boldsymbol{r}} - \frac{1}{\sqrt{2}}\left(\hat{Q} - i\hat{P}\right)e^{-i\boldsymbol{k}\cdot\boldsymbol{r}}\right) \tag{3.42}$$

or, using Eqs. (3.35) and (3.36):

$$\widehat{E}(r) = i\varepsilon\varepsilon^1 \left(\widehat{a}e^{ik\cdot r} - \widehat{a}^\dagger e^{-ik\cdot r}\right) \tag{3.43}$$

Note that the time-dependence is absent from Eqs. (3.42) and (3.43), which is discussed in the next section.

3.5 Time-Dependence

According to the "Schrodinger picture", the states evolve in time, while the operators are time-independent. In this viewpoint, the time-dependence of $\widehat{E}$ is absent, because $\widehat{E}$ is an operator:

$$\widehat{E}(r) = i\varepsilon\varepsilon^1 \left(\widehat{a}e^{ik\cdot r} - \widehat{a}^\dagger e^{-ik\cdot r}\right) \tag{3.44}$$

The time-dependence is contained in the state, not in the operator:

$$|\psi(t)\rangle = |\psi(0)\rangle \, e^{-i\omega t} \tag{3.45}$$

The physical observable is

$$\langle E(r, \ t)\rangle = \langle\psi(t)|\widehat{E}(r)|\psi(t)\rangle \tag{3.46}$$

Note that for stationary states $|\psi(t)\rangle = |\psi(0)\rangle \, e^{-i\omega t}$ while $\langle\psi(t)| = \langle\psi(0)|e^{+i\omega t}$, so the time-dependence in Eq. (3.46) cancels out.

Alternatively, according to the "Heisenberg picture", the operators are time-dependent while the states are time-independent and keep their initial value at some time t_0:

$$\widehat{E}(r, t) = i\varepsilon\varepsilon^1 \left(\widehat{a}e^{i(k\cdot r - \omega t)} - \widehat{a}^\dagger e^{-i(k\cdot r - \omega t)}\right) \tag{3.47}$$

The physical observable is

$$\langle E(r, \ t)\rangle = \langle\psi(t_0)|\widehat{E}(r, t)|\psi(t_0)\rangle \tag{3.48}$$

As we will see later, the Heisenberg viewpoint is useful in some circumstances, for example, when describing photodetection events at two different times.

3.6 Quadrature Operators

Repeating the results obtained above, we have the quantum description of the field:

$$
\begin{aligned}
\widehat{E}(\mathbf{r}) &= i\varepsilon\varepsilon^1 \left(\widehat{a}e^{i\mathbf{k}\cdot\mathbf{r}} - \widehat{a}^\dagger e^{-i\mathbf{k}\cdot\mathbf{r}} \right) \\
&= i\varepsilon\varepsilon^1 \left(\frac{1}{\sqrt{2}} \left(\widehat{Q} + i\widehat{P} \right) e^{i\mathbf{k}\cdot\mathbf{r}} - \frac{1}{\sqrt{2}} \left(\widehat{Q} - i\widehat{P} \right) e^{-i\mathbf{k}\cdot\mathbf{r}} \right) \\
&= \varepsilon\varepsilon^1 \sqrt{2} \left(-\widehat{Q}\frac{(e^{i\mathbf{k}\cdot\mathbf{r}} - e^{-i\mathbf{k}\cdot\mathbf{r}})}{2i} - \widehat{P}\frac{(e^{i\mathbf{k}\cdot\mathbf{r}} + e^{-i\mathbf{k}\cdot\mathbf{r}})}{2} \right) \\
&= -\varepsilon\varepsilon^1 \sqrt{2} \left[\widehat{Q} \sin (\mathbf{k}\cdot\mathbf{r}) + \widehat{P} \cos (\mathbf{k}\cdot\mathbf{r}) \right]
\end{aligned}
\tag{3.49}
$$

$\widehat{Q}$ and $\widehat{P}$ describe the two components of the field that are $\frac{\pi}{2}$ out of phase. Recall the definitions of $\widehat{Q}$ and $\widehat{P}$:

$$
\widehat{Q} = \frac{1}{\sqrt{2}} \left(\widehat{a} + \widehat{a}^\dagger \right)
\tag{3.50}
$$

$$
\widehat{P} = \frac{-i}{\sqrt{2}} \left(\widehat{a} - \widehat{a}^\dagger \right)
\tag{3.51}
$$

Using Eqs. (3.50) and (3.51), the expectation values (average) of $\widehat{Q}$ and $\widehat{P}$ in the QHO state, $|n\rangle$, are easily evaluated using the results from Chap. 2:

$$
\langle Q \rangle = \frac{1}{\sqrt{2}} \langle n|(\widehat{a} + \widehat{a}^\dagger)|n\rangle = 0
\tag{3.52}
$$

$$
\langle P \rangle = \frac{-i}{\sqrt{2}} \langle n|(\widehat{a} - \widehat{a}^\dagger)|n\rangle = 0
\tag{3.53}
$$

Exercise 3.3 Derive Eqs. (3.52) and (3.53).

The expectation value of Q^2 is

$$
\begin{aligned}
\langle Q^2 \rangle &= \frac{1}{2} \langle n|(\widehat{a} + \widehat{a}^\dagger)(\widehat{a} + \widehat{a}^\dagger)|n\rangle \\
&= \frac{1}{2} \langle n|(\widehat{a}\widehat{a} + \widehat{a}\widehat{a}^\dagger + \widehat{a}^\dagger\widehat{a} + \widehat{a}^\dagger\widehat{a}^\dagger)|n\rangle
\end{aligned}
\tag{3.54}
$$

Examining each term, we have $\langle n|\widehat{a}\widehat{a}|n\rangle = 0$ and $\langle n|\widehat{a}^\dagger\widehat{a}^\dagger|n\rangle = 0$, giving

$$\langle Q^2 \rangle = \frac{1}{2}\langle n|(\hat{a}\,\hat{a}^\dagger + \hat{a}^\dagger \hat{a})|n\rangle \qquad (3.55)$$

Using the commutation relation, $[\hat{a}, \hat{a}^\dagger] = 1$, gives

$$\langle Q^2 \rangle = \frac{1}{2}\langle n|(1 + 2\hat{a}^\dagger \hat{a})|n\rangle$$
$$= n + \frac{1}{2} \qquad (3.56)$$

Similarly,

$$\langle P^2 \rangle = n + \frac{1}{2} \qquad (3.57)$$

The uncertainties are

$$\Delta Q = \sqrt{\langle Q^2 \rangle - \langle Q \rangle^2} = \sqrt{n + \frac{1}{2}} \qquad (3.58)$$

$$\Delta P = \sqrt{\langle P^2 \rangle - \langle P \rangle^2} = \sqrt{n + \frac{1}{2}} \qquad (3.59)$$

which is the same as Eqs. (2.115) and (2.116).

The product of uncertainties is

$$\Delta Q \Delta P = n + \frac{1}{2} \qquad (3.60)$$

which is identical to Eq. (2.117). According to Eq. (3.60), the state $|0\rangle$ with $n = 0$ is the minimum uncertainty state with $\Delta Q \Delta P = \frac{1}{2}$. Thus,

$$\Delta Q \Delta P \geq \frac{1}{2} \qquad (3.61)$$

which is identical to Eq. (2.118). Equation (3.61) is known as the standard quantum limit (SQL).

We can represent $\langle Q \rangle$ and $\langle P \rangle$ in Fig. 3.5. According to Eqs. (3.52) and (3.53), Q and P are centred at the origin in Fig. 3.5, because their average is zero. However, according to Eqs. (3.58) and (3.59), there is a spread (uncertainty) in Q and P, represented by the gray circle with area approximated by $\Delta Q \Delta P$, in accordance with Eq. (3.61). Note, however, that the distribution represented by the gray circle is actually a Gaussian distribution according to Eq. (2.57) for the state $|0\rangle$. Q and P do not commute and cannot be measured simultaneously, so a series of different measurements are needed to determine Q and P. The circle represents what would be obtained after plotting many measurements of Q and P.

Fig. 3.5 Quadrature
representation of the
quantum harmonic oscillator
in state $|0\rangle$

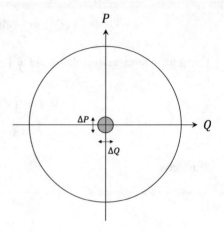

3.7   Physical Observables

As mentioned in Chap. 2, $\widehat{a}$ and $\widehat{a}^{\dagger}$ are not Hermitian, meaning they do not
correspond to any physical observable. However, the field $\widehat{E}$, number operator $\widehat{N}$,
Hamiltonian $\widehat{H}$, and quadratures $\widehat{Q}$ and $\widehat{P}$ are Hermitian, although $\widehat{a}$ and $\widehat{a}^{\dagger}$ are not.
Thus, $\widehat{E}$, $\widehat{N}$, $\widehat{H}$, $\widehat{Q}$ and $\widehat{P}$ correspond to physical observables. For example, at low
frequencies (e.g., radio), antennas can directly detect the electric field. Photodetec-
tors measure light intensity proportional to the number of photons, given by the
number operator $\widehat{N} = \widehat{a}^{\dagger}\widehat{a}$ (photon-resolving detectors also exist that can determine
the number of photons in a state). Bolometers measure field energy from a rise in
temperature associated with the Hamiltonian, $\widehat{H}$. In Chap. 12, we will see that $\widehat{Q}$ and
$\widehat{P}$ can be measured by a technique called homodyne detection.

Exercise 3.4 Show that $\widehat{E}$, $\widehat{N}$, $\widehat{H}$, $\widehat{Q}$ and $\widehat{P}$ are Hermitian.

Another Hermitian operator is the momentum, $\widehat{p}$. The momentum operator can be
written as

$$\widehat{p} = \hbar k \widehat{a}^{\dagger}\widehat{a} \tag{3.62}$$

$\widehat{p}$ applied to the state, $|n\rangle$, gives

$$\begin{aligned} \widehat{p}|n\rangle &= \hbar k \widehat{a}^{\dagger}\widehat{a}|n\rangle \\ &= \hbar k n|n\rangle \end{aligned} \tag{3.63}$$

Thus, $|n\rangle$ is an eigenstate of $\widehat{p}$ with eigenvalue $n\hbar k$.

Fig. 3.6 Gilbert Lewis. (Credit: AIP Emilio Segrè Visual Archives, photograph by Francis Simon)

3.8 Photons

The concept of photon emerges from the Dirac formalism applied to the electromagnetic field. $|n\rangle$ corresponds to a state with energy $n\hbar\omega$ above the vacuum energy and momentum $n\hbar k$. Thus, a photon is an elementary excitation of the quantized electromagnetic field that resembles a particle with energy $\hbar\omega$ and momentum $\hbar k$. The state $|n\rangle$ contains n photons. The annihilation operator, $\hat{a}$, destroys a photon and the creation operator, $\hat{a}^\dagger$, creates a photon. Gilbert Lewis (Fig. 3.6), an American chemist, coined the term "photon" in 1926 [7]. In general, a quantized field allows particle creation and annihilation, and is the foundation for quantum field theory.

References

1. A. Einstein, Ann. Phys., Lpz. 17 (1905) 132.
2. A. Einstein, Ann. Phys., Lpz. 20 (1906) 199.
3. A. Einstein, Ann. Phys., Lpz. 22 (1907) 180.
4. A. Einstein, Phys. Z. 10 (1909) 185.
5. A. Einstein, Phys. Z. 10 (1909) 817.
6. File: Einstein 1921 by F Schmutzer – restoration.jpg. (2020, September 26). *Wikimedia Commons, the free media repository*. Retrieved 10:26, December 7, 2020 from https://commons.wikimedia.org/w/index.php?title=File:Einstein_1921_by_F_Schmutzer_-_restoration.jpg&oldid=471895083
7. G.N. Lewis, *The conservation of photons*, Nature 118 (1926) 874.

Chapter 4
Fock States and the Vacuum

In this chapter, we further examine the properties of the Fock or number state, $|n\rangle$. The average photon number and average electric field of a Fock state are derived, along with their uncertainty. We show that the Fock state has the unusual property of having a precise number of photons, n, but with average electric field of zero. Although the average field is zero, there is an uncertainty or fluctuation in the field. The properties of the Fock state $|0\rangle$, called the vacuum, are presented along with the concept of vacuum field fluctuations and their experimental consequences.

4.1 Photon Number

The average number of photons (expectation value of n) in Fock state $|n\rangle$ is

$$\langle n \rangle = \langle n|\widehat{N}|n\rangle = \langle n|\widehat{a}^\dagger\widehat{a}|n\rangle = n \tag{4.1}$$

The expectation value of n^2 is obtained by applying the $\widehat{N}$ operator twice:

$$\langle n^2 \rangle = \langle n|\widehat{N}^2|n\rangle = n\langle n|\widehat{N}|n\rangle = n^2 \tag{4.2}$$

The uncertainty in n is

$$\Delta n = \sqrt{\langle n^2 \rangle - \langle n \rangle^2} = \sqrt{n^2 - (n)^2} = 0 \tag{4.3}$$

Thus, the Fock state $|n\rangle$ has a definite number of photons, n, with zero uncertainty.

© The Author(s), under exclusive license to Springer Nature Switzerland AG 2022
R. LaPierre, *Getting Started in Quantum Optics*, Undergraduate Texts in Physics,
https://doi.org/10.1007/978-3-031-12432-7_4

Exercise 4.1 We could write $\widehat{N}^2$ as $\hat{a}^\dagger \hat{a} \hat{a}^\dagger \hat{a}$. Show that if we put $\hat{a}^\dagger \hat{a} \hat{a}^\dagger \hat{a}$ in the normal order, we get $\langle n | \hat{a}^\dagger \hat{a} \hat{a}^\dagger \hat{a} | n \rangle = n^2$, the same as Eq. (4.2).

4.2   Electric Field of the Fock State

Recall from Chap. 3 that the electric field operator is

$$\widehat{E}(r) = i\varepsilon\varepsilon^1 (\hat{a} e^{ik \cdot r} - \hat{a}^\dagger e^{-ik \cdot r}) \tag{4.4}$$

The average (expectation value) of $\widehat{E}$ in the state $|n\rangle$ is

$$\begin{aligned}
\langle E \rangle &= \langle n | \widehat{E} | n \rangle \\
&= i\varepsilon\varepsilon^1 \langle n | (\hat{a} e^{ik \cdot r} - \hat{a}^\dagger e^{-ik \cdot r}) | n \rangle \\
&= 0
\end{aligned} \tag{4.5}$$

since $\hat{a} |n\rangle = \sqrt{n} \, |n-1\rangle$, and $\langle n | n-1 \rangle = 0$ because these are orthogonal states. Similarly, $\langle n | \hat{a}^\dagger | n \rangle = 0$. Thus, we have the unusual situation where the Fock state, $|n\rangle$, has a precise number of photons, n, but the average electric field is zero. This tells us that Fock states are a nonclassical form of light. Due to their nonclassical nature, Fock states are very difficult to produce. In the next chapter, we will examine the single photon state, that is, the Fock state with $n = 1$, including methods to produce this state.

Although the average electric field of a Fock state is zero, the field can still fluctuate about the average. The average of the square of the field (expectation value of $\widehat{E}^2$) is

$$\begin{aligned}
\langle E^2 \rangle &= \langle n | \widehat{E}^2 | n \rangle \\
&= (i*i)(\varepsilon^1)^2 \langle n | (\hat{a} e^{ik \cdot r} - \hat{a}^\dagger e^{-ik \cdot r})(\hat{a} e^{ik \cdot r} - \hat{a}^\dagger e^{-ik \cdot r}) | n \rangle \\
&= -(\varepsilon^1)^2 \langle n | (\hat{a}^2 e^{2ik \cdot r} - \hat{a} \hat{a}^\dagger - \hat{a}^\dagger \hat{a} + (\hat{a}^\dagger)^2 e^{-2ik \cdot r}) | n \rangle
\end{aligned} \tag{4.6}$$

Keeping only the non-zero terms in Eq. (4.6), we get

$$\langle E^2 \rangle = (\varepsilon^1)^2 \langle n | (\hat{a} \hat{a}^\dagger + \hat{a}^\dagger \hat{a}) | n \rangle \tag{4.7}$$

Putting Eq. (4.7) into the normal order (or, alternatively, using Eqs. (2.127) and (2.128)) gives

$$\langle E^2 \rangle = \left(\varepsilon^1\right)^2 (2n+1) \tag{4.8}$$

The field fluctuations are described by the uncertainty:

$$\Delta E = \sqrt{\langle E^2 \rangle - \langle E \rangle^2}$$
$$= \varepsilon^1 \sqrt{2n+1} \tag{4.9}$$

Of course, there are corresponding fluctuations in the magnetic field too. Assuming we could repeat a field measurement multiple times on identically prepared Fock states, we would obtain a distribution in the measurement result according to Eq. (4.9), and an average of zero according to Eq. (4.5). Equation (4.9) tells us that the field fluctuations increase as the number of photons increases.

4.3 Vacuum Fluctuations

Classically, we think of the vacuum as an absence of everything. Quantum mechanically, this is not true. For $n = 0$, that is, the vacuum state $|0\rangle$, we have $\Delta E = \varepsilon^1$. Thus, $|0\rangle$ is a minimum uncertainty state with the least uncertainty in the electric field. The $|0\rangle$ state also corresponds to the minimum uncertainty of the quadratures where $\Delta Q \Delta P = \frac{1}{2}$ with equal uncertainty in Q and P. We can summarize the vacuum properties as follows:

$$|0\rangle : \text{vacuum state with } n = 0 \text{ photons} \tag{4.10}$$

$$\langle n \rangle = \langle 0 | \widehat{N} | 0 \rangle = 0 \tag{4.11}$$

$$\text{Vacuum energy} : E_0 = \hbar\omega\left(n + \frac{1}{2}\right) = \hbar\omega\left(0 + \frac{1}{2}\right) = \frac{1}{2}\hbar\omega \tag{4.12}$$

$$\text{Average electric field } \langle E \rangle = \langle 0 | \widehat{E} | 0 \rangle = 0 \tag{4.13}$$

$$\text{Field fluctuations} : \Delta E = \varepsilon^1 \tag{4.14}$$

$$\Delta Q \Delta P = \frac{1}{2}, \text{minimum uncertainty state} \tag{4.15}$$

There are no photons in the vacuum and, on average, there is no electric field in the vacuum either. However, there are fluctuations, ε^1, in the field. These field fluctuations have quadrature values at the minimum allowed by the uncertainty relation, $\Delta Q \Delta P = \frac{1}{2}$. We can think of "empty" space as filled with these field fluctuations.

4.4 Experimental Evidence of Vacuum Fluctuations

Are vacuum fluctuations real or are they just a figment of the physicist's imagination? The experimental consequence of $\langle n \rangle = 0$ and $\langle E \rangle = 0$ for the vacuum state is that we cannot directly photodetect the vacuum. However, we have an abundance of indirect evidence for vacuum fluctuations, including spontaneous emission, the Lamb shift, the magnetic moment of the electron, the Casimir effect, and the van der Waals force. We consider each of these below.

In quantum mechanics, an energy eigenstate of the Hamiltonian is a stationary state. This means that an electron in an excited state (an eigenstate of the Hamiltonian) should remain there forever, in the absence of any external perturbations. Spontaneous emission, involving the transition of an electron from an excited to the ground state, is explained by the perturbation on the electron from the field fluctuations of the vacuum. Aspects of spontaneous emission are described in Chaps. 9 and 19.

The Lamb shift, named after Willis Lamb (Fig. 4.1a), is a correction to certain energy levels of the hydrogen atom. This effect is due to the perturbation of the electron by the field fluctuations of the vacuum. The electric field fluctuations of the vacuum cause a rapid oscillation of the electron in the Coulomb potential of the hydrogen atom, causing a slight shift in the energy levels of the s orbitals. A

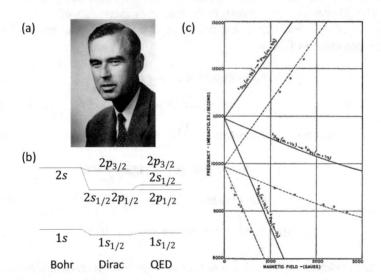

Fig. 4.1 (a) Willis Lamb (Nobel Prize in Physics in 1955). (Credit: AIP Emilio Segrè Visual Archives, W. F. Meggers Gallery of Nobel Laureates Collection). (b) Energy levels according to Bohr, Dirac, and quantum electrodynamics (QED). The Lamb shift is the increase of the $2s_{1/2}$ level according to QED. (c) Original experimental data of the Lamb shift by Lamb and Retherford. The energy levels vary in a magnetic field due to the Zeeman effect. The experimental data (circles) are shifted down by 1000 Megacycles for easier comparison with the theory (solid lines). (Reprinted with permission from Lamb Jr. and Retherford [1], https://doi.org/10.1103/PhysRev.72.241. Copyright 1947 by the American Physical Society)

derivation of the Lamb shift is given in Appendix 1. The energy levels of the hydrogen atom according to Bohr, Dirac, and quantum electrodynamics (QED) are shown in Fig. 4.1b. According to the simple hydrogenic model by Bohr, the energy levels of the hydrogen atom depend only on the principal quantum number, n, and not on the orbital angular momentum quantum number l ($l = 0$ for the s orbital and $l = 1$ for the p orbital). Dirac extended the Bohr theory to include the effects of electron spin (spin–orbit coupling). In Dirac theory, the $2s_{1/2}$ ($n = 2$, $l = 0$, $j = \frac{1}{2}$) and $2p_{1/2}$ ($n = 2$, $l = 1$, $j = \frac{1}{2}$) energy levels are degenerate (the subscript indicates the total angular momentum quantum number, j, which includes orbital angular momentum and electron spin). Finally, QED includes the effect of the vacuum field fluctuations. The increase of the $2s_{1/2}$ energy level predicted by QED is called the Lamb shift, which removes the degeneracy between the $2s_{1/2}$ and $2p_{1/2}$ energy levels. The energy splitting is about 1058 MHz, in the microwave frequency range. The Lamb shift in the hydrogen spectrum was first measured in 1947 by Lamb and Retherford (Fig. 4.1c) [1] using microwave resonance. In 1947, Hans Bethe explained the Lamb shift in terms of the vacuum fluctuations [2]. These experimental and theoretical efforts laid the foundation for QED.

The magnetic moment associated with the spin of the electron is given by $\mu = -g\mu_B m_s$ where μ_B is the Bohr magneton and m_s is the spin quantum number ($m_s = \pm\frac{1}{2}$ for the electron). g, called the "spin g-factor", is a correction factor predicted to have a value of 2 from the relativistic Dirac equation. However, the spin g-factor has been measured and has the actual value of 2.00231930436182. The small deviation from 2 is due to the effect of vacuum fluctuations on the electron magnetic moment, which can be calculated in QED. The theoretical prediction agrees with the experimentally measured value to about 1 part in a billion, making QED one of the most accurate physical theories that exists!

The vacuum energy can even have an influence on macroscopic systems. The Casimir effect, introduced in 1948 by Hendrik Casimir (Fig. 4.2a) [3], is a physical force associated with the vacuum field fluctuations. If two parallel neutral conducting plates are held closely together in vacuum, a force (called the Casimir force) pushes the two plates together. In classical electromagnetism, the force between neutral plates is zero. The force arises quantum mechanically due to a restriction in the allowed mode frequencies of the vacuum state between the plates due to the electromagnetic boundary conditions imposed by the plates. The electric field at the conducting plates must go to zero, according to Maxwell's equations for a perfect conductor. Thus, only certain discrete wavelengths can exist between the plates, as shown in Fig. 4.2b. Specifically, the wavevector perpendicular to the plate surface can take only certain discrete values of $k_z = n\pi/d$ where $n = 1, 2, 3, \ldots$, corresponding to standing waves between the plates. Therefore, there is a reduction in the number of allowed vacuum modes between the plates. As the separation between the plates increases, the number of allowed modes between the plates also increases, and hence, the vacuum energy between the plates increases. This tells us that the energy is minimized when the plates have zero separation, meaning there is a mutual attraction between the two plates. The difference in zero-point energy, ΔE, for the continuum of modes versus those between the plates can be calculated, and then the Casimir force per unit area of the plates as a function of plate separation

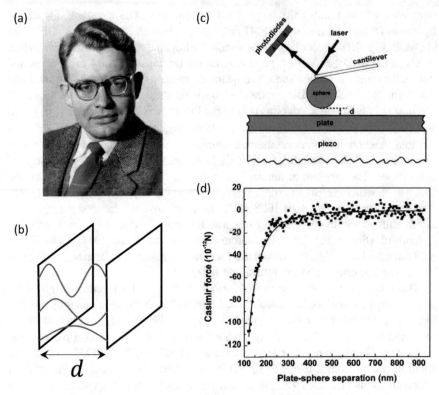

Fig. 4.2 (a) Hendrik Casimir. (Credit: AIP Emilio Segrè Visual Archives, Physics Today Collection). (b) Discretization of the vacuum mode wavelengths between parallel conducting plates leading to the Casimir effect. (c) Experimental method of measuring the Casimir force using a plate–sphere geometry [8]. (d) Experimental data of Casimir force versus plate–sphere separation. The solid line is the theoretical prediction [8]. ((c) and (d) are reprinted with permission from Mohideen and Roy [8], https://doi.org/10.1103/PhysRevLett.81.4549. Copyright 1998 by the American Physical Society)

d can be determined from $F = -\frac{\partial \Delta E}{\partial d}$. An equivalent view is that the radiation pressure from the vacuum is greater outside the plates than between the plates, causing a force that pushes the plates together. An analogy is the attractive force between two closely spaced ships at sea due to the restricted standing waves between them [4]. The resulting attractive force per unit area of the plates falls off sharply with separation according to [3]

$$F = -\frac{\pi^2 \hbar c}{240} \frac{1}{d^4} \qquad (4.16)$$

A simple heuristic derivation of the $1/d^4$ force dependence is given in Appendix 2. If $d = 1$ μm, then the Casimir force is only 10^{-7} N/cm^2. However, the force rises rapidly with decreasing plate separation. Thus, if $d = 1$ nm, then the force is 10^5 N/cm^2. Casimir stated in his 1948 publication [3], "although the effect is small, an

(a)   (b)

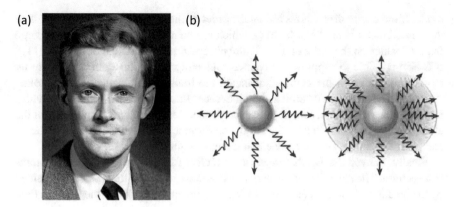

Fig. 4.3 (**a**) Edward Mills Purcell (Nobel Prize in Physics in 1952). (Credit: Wikimedia Commons [13]). (**b**) Enhancement in spontaneous emission from an atom in a cavity (right) compared to free space (left). (Reprinted by permission from Springer Nature, Vahala [14]. Copyright 2003)

experimental confirmation seems not unfeasible and might be of a certain interest." A decade later, in 1958, M.J. Sparnaay experimentally measured the force between two plates in qualitative agreement with Casimir's prediction [5]. Nearly 50 years after Casimir's prediction, S.K. Lamoreaux [6, 7] and others [8], beginning in 1997, used modern experimental techniques such as piezoelectric transducers to control the plate positions very accurately. The plates were also replaced with more convenient geometries, such as a plate and a sphere (Fig. 4.2c), since it is difficult to keep two large plates parallel with small separation. These experiments confirmed the theory to about 1% accuracy (Fig. 4.2d). Thus, the Casimir force is more than just a theoretical curiosity. It is a real effect in microelectromechanical systems and nanotechnology [9].

The Purcell effect, similar in some respects to the Casimir effect, is an enhancement of the spontaneous emission from an atom placed inside a resonant cavity. In the 1940s, Edward Mills Purcell (Fig. 4.3) discovered the effect named after him [10]. The cavity modifies the density of modes at the transition frequency of the atom, thereby modifying the spontaneous emission rate of the atom. An enhancement of the spontaneous emission rate is observed if the standing electric field in the cavity is resonant with the electronic transition energy of the atom. The enhancement factor, called the Purcell factor, is the ratio of the mode density in the cavity to that in vacuum and is given by [10]

$$F = \frac{3Q\lambda_o^{\,3}}{4\pi^2 V} \tag{4.17}$$

where Q is the quality factor of the cavity, λ_o is the wavelength in the cavity, and V is the mode volume inside the cavity. The quality factor is a measure of the "sharpness" of the cavity mode and is given by $Q = \lambda_o/\Delta\lambda$, where $\Delta\lambda$ is the linewidth of the cavity mode. If the cavity dimensions are on the order of the field wavelength, $V \sim \lambda_o^3$, then the Purcell factor is on the order of Q, which can be many orders of magnitude large (e.g., Q values on the order of 10–100 are observed in semiconductor cavities [11]). It is also possible to suppress the spontaneous emission of the atom if there are no mode frequencies in the cavity that match the transition frequency of the atomic emission. The atom cannot emit a photon because the cavity contains no mode with a matching frequency to accept it. Thus, spontaneous emission is not a property of the atom, but can be controlled by the atom's environment. This is of both fundamental and practical significance for the engineering of light sources [12].

Finally, the van der Waals force is an attractive force between two neutral atoms in a vacuum. The electrostatic attraction between two neutral atoms can be explained by the creation of induced electric dipole moments in the atoms by the vacuum field fluctuations.

Exercise 4.2 The Casimir effect is a force produced from vacuum fluctuations. Could free energy be harnessed from the vacuum?

References

1. W.E. Lamb, Jr. and R.C. Retherford, *Fine structure of the hydrogen atom by a microwave method*, Phys. Rev. 72 (1947) 241.
2. H.A. Bethe, *The electromagnetic shift of energy levels*, Phys. Rev. 72 (1947) 339.
3. H.B.G. Casimir, *On the attraction between two perfectly conducting plates*, Proc. Kon. Ned. Akad. Wet. 51 (1948) 793.
4. S. L. Boersma, *A maritime analogy of the Casimir effect*, Am. J. Phys. 64 (1996) 539.
5. M.J. Sparnaay, *Measurements of attractive forces between flat plates*, Physica 24 (1958) 751.
6. S.K. Lamoreaux, *Demonstration of the Casimir force in the 0.6–6µm range*, Phys. Rev. Lett. 78 (1997) 5.
7. S.K. Lamoreaux, *The Casimir force: background, experiments, and applications*, Rep. Prog. Phys. 68 (2004) 201.
8. U. Mohideen and A. Roy, *Precision measurement of the Casimir force from 0.1 to 0.9 µm*, Phys. Rev. Lett. 81 (1998) 4549.
9. F. Capasso, J.N. Munday, D. Iannuzzi and H.B. Chan, *Casimir forces and quantum electrodynamical torques: Physics and nanomechanics*, IEEE J. Selected Topics in Quantum Electronics 13 (2007) 400.
10. E.M. Purcell, *Spontaneous emission probabilities at radio frequencies*, Phys. Rev. 69 (1946) 681 (note B10).
11. J.-M. Gérard and B. Gayral, *Strong Purcell effect for InAs quantum boxes in three-dimensional solid-state microcavities*, J. Lightwave Technol. 17 (1999) 2089.
12. T.F. Krauss and R.M. De La Rue, *Photonic crystals in the optical regime – past, present and future*, Progress in Quantum Electronics 23 (1999) 51.
13. https://commons.wikimedia.org/wiki/File:Edward_Mills_Purcell.jpg
14. K.J. Vahala, *Optical microcavities*, Nature 424 (2003) 839.

Chapter 5
Single Photon State

In the previous chapter, we examined the properties of the special Fock state, $|0\rangle$. In this chapter, another special Fock state, $|1\rangle$, called the single photon state, is presented. The quantum optics treatment of photodetection is explained. Methods of generating and detecting single photons are described.

5.1 Single Photon State

Single photons are required for quantum communications, quantum computing, quantum metrology, and quantum information processing based on the optical approach. Using the results of Chap. 3, the single photon state $|1\rangle$ (Fock state with $n = 1$) has the following properties:

$$\langle n \rangle = 1 \tag{5.1}$$

$$\Delta n = 0 \tag{5.2}$$

$$E_1 = \frac{3}{2}\hbar\omega \tag{5.3}$$

$$\langle E \rangle = \langle 0|\widehat{E}|0\rangle = 0 \tag{5.4}$$

$$\Delta E = \sqrt{3}\varepsilon^1 \tag{5.5}$$

$$\Delta Q \Delta P = \frac{3}{2} \tag{5.6}$$

According to Eqs. (5.5) and (5.6), the single photon state is not a minimum uncertainty state. Note that the single photon energy, $\hbar\omega$, is the photon energy measured above the vacuum energy, $E_1 - E_0 = \hbar\omega$.

© The Author(s), under exclusive license to Springer Nature Switzerland AG 2022
R. LaPierre, *Getting Started in Quantum Optics*, Undergraduate Texts in Physics,
https://doi.org/10.1007/978-3-031-12432-7_5

5.2 Photodetection Signal

The direct detection of a single photon of a Fock state $|n\rangle$ destroys a single photon, represented by $\widehat{a}|n\rangle$. Intuitively, light intensity is expected to be proportional to the number of photons:

$$I \propto \langle n \rangle = \langle n|\widehat{a}^\dagger a|n\rangle = n \tag{5.7}$$

Alternatively, according to classical physics, intensity measured by a photodetector is proportional to the square of the electric field ($I \propto E^2$). Quantum mechanically, we can write the electric field as

$$\widehat{E}(r) = i\varepsilon\varepsilon^1\widehat{a}e^{ik\cdot r} + h.c. = \widehat{E}^+(r) + \widehat{E}^-(r) \tag{5.8}$$

where $h.c.$ is the Hermitian conjugate. $\widehat{E}^+(r) = i\varepsilon\varepsilon^1\widehat{a}e^{ik\cdot r}$ is called the positive frequency component of the field, and $\widehat{E}^-(r) = -i\varepsilon\varepsilon^1\widehat{a}^\dagger e^{-ik\cdot r}$ is called the negative frequency component of the field. The positive frequency component contains the annihilation operator, and is thus responsible for photon absorption, while the negative frequency component contains the creation operator and is responsible for emission of a photon. Since we expect intensity measurements at a photodetector to be related to absorption, we can write

$$I \propto \left\| \widehat{E}^+(r)|n\rangle \right\|^2 \tag{5.9}$$

$$= \langle n|\widehat{E}^-(r)\widehat{E}^+(r)|n\rangle \tag{5.10}$$

$$= \left(\varepsilon^1\right)^2 \langle n|\widehat{a}^\dagger\widehat{a}|n\rangle \tag{5.11}$$

$$= \left(\varepsilon^1\right)^2 n \tag{5.12}$$

Note that the intensity is proportional to $\langle n|\widehat{a}^\dagger\widehat{a}|n\rangle$, which is equal to the photon number as we expect intuitively from Eq. (5.7). By setting $n = 1$ in Eq. (5.12), $(\varepsilon^1)^2$ can be interpreted as the "intensity of a single photon", and ε^1 is the "field amplitude of a single photon". A word of caution is warranted, however, because the interpretation of ε^1 as the "field amplitude of a single photon" is not quite correct, since $\langle E \rangle = 0$ for the single photon state.

Suppose we start with a single photon state, $|1\rangle$. Direct photodetection destroys the photon, and the state becomes

$$\widehat{a}|1\rangle = |0\rangle \tag{5.13}$$

consistent with Eq. (2.127). If we measure again, the state becomes

$$\hat{a}|0\rangle = 0 \tag{5.14}$$

according to Eq. (2.129). By definition, you cannot detect a single photon twice, since the first photodetection event destroys the single photon.

Note that we are speaking above about direct detection of a photon, which destroys the photon by absorption. It is sometimes desirable to perform a measurement of a photon without destroying the photon; that is, we want to make a measurement of the photon without disturbing it and then let it continue on its way. This measurement process is known as "quantum non-demolition" (QND) measurement. QND usually involves forming an entangled state of the particle to be measured with another particle called the "meter particle" (entanglement is discussed in Chap. 8). A measurement of the meter particle is performed, which tells us the state of the measurement particle via the entanglement. The interested reader may obtain more information on QND in Refs. [1, 2].

5.3   Single Photon Sources

The ideal single photon source will produce single photons deterministically (on demand) with high collection efficiency, at a high repetition rate (GHz), at room temperature, at the desired wavelength (or spectrum), and with definite polarization. Ideally, the emitter can be triggered to emit single photons on demand by optical or electrical excitation. The photons should be produced with a high quantum yield (ideally 100%), meaning that each trigger event results in a single photon emission with certainty. Each photon wavepacket should be created in the identical state, so that they are indistinguishable from each other (i.e., the photons should have identical spectra, spatial distribution, and polarization); hence, they can interfere perfectly as required for quantum information processing applications. The indistinguishability of photons can be tested through a Hong–Ou–Mandel experiment, described in Chap. 17. The source should produce single photons of high purity, meaning that two-photon correlations are negligible (see Chap. 6 for photon correlations and their measurement). Thus far, there is no single photon source that meets all the above criteria perfectly, and the development of improved single photon sources remains an active research topic.

The two main types of single photon source available today are radiative decay in two-level systems or spontaneous parametric down-conversion (SPDC) in nonlinear crystals [3]. In two-level systems, spontaneous emission from an excited state to the ground state creates a single photon wavepacket (wavepackets are discussed in Chap. 9). Two-level systems include atoms, ions (Fig. 5.1a), nitrogen-vacancy (NV) centers in diamond (Fig. 5.1b) or other color centers, semiconductor quantum dot heterostructures (Fig. 5.1c, d), and semiconductor nanocrystals (Fig. 5.1e).

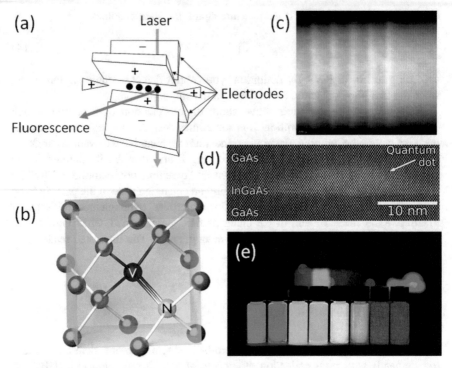

Fig. 5.1 Single photon sources. (**a**) Atomic or ionic emitters (represented by the circles). Electrodes are used to confine the ions, and lasers are used to control the state of the ions and read out their state by fluorescence emission. (**b**) Illustration of the crystal structure of the N-V center in diamond (green atoms are carbon), containing a substitutional nitrogen atom and a carbon vacancy. (**c**) Transmission electron microscopy image of GaAs quantum dots (QDs) formed along the length of a GaP nanowire. Nanowire diameter is about 50 nm. (**d**) Cross-sectional transmission electron microscopy image of an InGaAs quantum dot in a GaAs matrix. (**e**) Colloidal quantum dots showing size-dependent fluorescence under ultraviolet light. (**b**, **d**, **e**). (Credit: Wikimedia Commons [4])

Quantum dots (QDs) are structures in which electron motion is strongly confined in all three dimensions (with energy quantization in all directions), leading to discrete energy levels. Today, QDs may be realized by a variety of different techniques and in a range of different materials (Fig. 5.1c–e). For example, QDs can be formed along the length of a nanowire by epitaxial growth methods (Fig. 5.1c), as a two-dimensional array of QDs on a surface using the Stranski–Krastanov process (Fig. 5.1d), or as nanocrystals in colloidal solutions (Fig. 5.1e).

Spontaneous parametric down-conversion (SPDC) is a nonlinear process that can occur in certain crystals such as potassium dihydrogenphosphate (KDP), beta barium borate (BBO), or lithium niobate (LiNbO$_3$). In SPDC, a pump laser occasionally undergoes a nonlinear interaction with the crystal. This interaction produces two lower-frequency photons (ω_s, ω_i), called the signal and idler, from the higher-frequency photon (ω_p) of the incident pump laser (Fig. 5.2). The frequencies and

Fig. 5.2 (a) Spontaneous
parametric down-
conversion (SPDC) in a
nonlinear crystal. (b) Energy
diagram of the SPDC
process

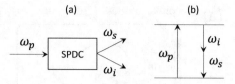

wavevectors of the three photons satisfy the conservation of energy and momentum
(the latter is called the phase matching condition):

$$\omega_p = \omega_s + \omega_i \tag{5.15}$$

$$k_p = k_s + k_i \tag{5.16}$$

The SPDC process is called degenerate if the down-converted photons have the same
frequency ($\omega_s = \omega_i = \omega_p/2$) and is called nondegenerate otherwise. In general, the
photons leaving the crystal propagate in different directions as illustrated in
Fig. 5.2a. Thus, the idler photon can be used as a trigger at a detector to announce
("herald") the arrival of the signal photon at another detector. For this reason, SPDC
is called a heralded single photon source. In addition, the photon pairs along certain
directions can be polarization entangled (entanglement is discussed in Chap. 8). A
drawback of the SPDC process is that the photons are produced by a probabilistic
process, rather than being produced on demand.

5.4   Single Photon Detectors

Remarkably, the human retina can detect single photons [5]. More practical single
photon detectors include the single photon avalanche diode (SPAD), the
photomultiplier tube (PMT), the superconducting nanowire single photon detector
(SNSPD), the electron multiplying charge-coupled device (EMCCD) camera, and
the intensified CCD (ICCD) camera [3]. Single photon detectors should have a high
detection efficiency and ideally be able to count individual photons.

A photomultiplier tube (PMT) (Fig. 5.3a) uses a photocathode to convert incident
photons into electrons. The electrons are subsequently amplified by secondary
emission at a series of dynodes resulting in electron multiplication and eventual
collection at an anode. PMTs have largely been replaced by semiconductor
photodiodes.

A single photon avalanche diode (SPAD) (Fig. 5.3b) uses the avalanche multiplication process in a silicon or III-V semiconductor p-n junction to convert an
incident photon into an electrical pulse.

The electron multiplying charge-coupled device (EMCCD) camera (Fig. 5.3c)
uses the avalanche process in an electron multiplying register of a charge-coupled
device (CCD) camera to amplify the CCD image [6]. The intensified CCD (ICCD)
camera (Fig. 5.3e) uses an image intensifier mounted in front of a CCD. The

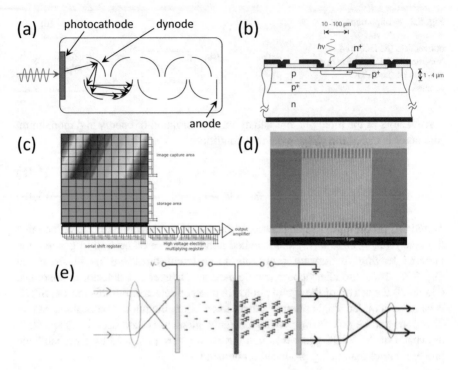

Fig. 5.3 Single photon detectors. (**a**) Photomultiplier tube (PMT). (**b**) Cross-sectional illustration of the p-n junctions in a single photon avalanche diode (SPAD). (**c**) Electron multiplying charge-coupled device (EMCCD) camera. (**d**) Superconducting nanowire single photon detector (SNSPD), showing the meandering superconducting nanowire. (**e**) Intensified CCD (ICCD) camera, showing photocathode (gray), microchannel plate (red), and phosphor (green). (**b–e**). (Credit: Wikimedia Commons [7])

intensifier consists of a photocathode to convert incident photons into electrons (similar to the PMT). The electrons are amplified using a microchannel plate, basically a microscopic version of the dynodes in a PMT. The electrons are then converted by a phosphor back to photons for detection by a CCD. EMCCD and ICCD cameras are used in quantum imaging (Chap. 18).

The superconducting nanowire single photon detector (SNSPD) (Fig. 5.3d) uses the photon energy to break superconducting Cooper pairs, creating a resistive state in a current-biased superconducting nanowire. The photon is detected as a voltage pulse due to the change in resistance.

Exercise 5.1 Investigate and explain the working principles in more detail for single photon sources and detectors.

References

1. P. Grangier, J.A. Levenson and J.-P. Poizat, *Quantum non-demolition measurements in optics*, Nature 396 (1998) 537.
2. G. Nogues et al., *Seeing a single photon without destroying it*, Nature 400 (1999) 239.
3. M.D. Eisaman, J. Fan, A. Migdall and S.V. Polyakov, *Single-photon sources and detectors*, Rev. Sci. Instrum. 82 (2011) 071101.
4. https://commons.wikimedia.org/wiki/File:Nitrogen-vacancy_center.png; https://commons.wikimedia.org/wiki/File:Gaas_inas_quantum_dot.jpg; https://commons.wikimedia.org/wiki/File:Quantum_Dots_with_emission_maxima_in_a_10-nm_step_are_being_produced_at_PlasmaChem_in_a_kg_scale.jpg;
5. J.N. Tinsley et al., *Direct detection of a single photon by humans*, Nat. Commun. 7 (2016) 12172.
6. L. Zhang et al., *A characterization of the single-photon sensitivity of an electron multiplying charge-coupled device*, J. Phys. B: At. Mol. Opt. Phys. 42 (2009) 114011.
7. https://commons.wikimedia.org/wiki/File:SPAD_Cross-section.gif; https://commons.wikimedia.org/wiki/File:EMCCD2_color_en.svg; https://en.wikipedia.org/wiki/File:NIST_SEM_Image_of_Superconducting_Nanowire_Single_Photon_Detector.jpg; https://commons.wikimedia.org/wiki/File:Image_intensifier_diagram.png

Chapter 6
Single Photon on a Beam Splitter

The beam splitter is an important optical element in quantum optics experiments. The classical and quantum treatment of the beam splitter is presented. We derive the photodetection probabilities for a single photon on a beam splitter, including single photon detection and double photon detection (coincidence counting). The correlation function is introduced for classical and quantum light. We show that the beam splitter creates an entangled state from a single photon input. The Hanbury Brown–Twiss experiment is introduced for characterizing light sources.

6.1 Classical Beam Splitter

The beam splitter is an important optical element in both classical and quantum optics experiments. As shown in Fig. 6.1, the beam splitter contains two input ports (labelled 1 and 2) and two output ports (labelled 3 and 4), also called modes. The beam splitter contains an interface that splits the incident electric field into a reflected and transmitted field. The reflection coefficient, r, is the fraction of incident field that is reflected. The transmission coefficient, t, is the fraction of incident field that is transmitted. r and t are complex numbers, which describe the magnitude and phase change of the incident field upon reflection and transmission.

For example, consider an electric field incident from a medium of low refractive index onto a medium of high refractive index (this case is called "external reflection"). According to the Fresnel equations familiar from classical optics [1], the reflected field in this situation will undergo a π phase shift (corresponding to a change in field amplitude of $e^{i\pi} = -1$) and the transmitted field will have zero phase shift. Conversely, if the field is incident from the opposite direction (from high to low refractive index, called "internal reflection"), then neither the reflected nor the transmitted field has any phase change. Thus, the beam splitter gives a π phase shift upon reflection from one direction only. This situation can arise in a beam splitter composed of a semitransparent metallic film on a glass substrate, for example.

© The Author(s), under exclusive license to Springer Nature Switzerland AG 2022
R. LaPierre, *Getting Started in Quantum Optics*, Undergraduate Texts in Physics,
https://doi.org/10.1007/978-3-031-12432-7_6

Fig. 6.1 A beam splitter
with input fields, E_1 and E_2,
and output fields, E_3 and E_4,
measured by detectors D_3
and D_4. The dot indicates
the side with a π phase shift
of the reflected field

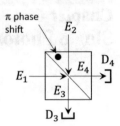

Throughout this book, a "dot" as in Fig. 6.1, will indicate the side of the beam splitter
that results in a π phase shift for the reflected field. It should be noted, however, that
the phase convention can vary [2, 3] depending on the technical design of the beam
splitter, and nowadays, beam splitters are usually composed of dielectric (not
metallic) films. A common phase convention is to adopt either a π phase shift in
one path of the beam splitter, or an *i* phase shift in *both* paths of the beam splitter.
Different phase conventions do not change our conclusions. Here, we adopt the π
phase shift convention.

As shown in Fig. 6.1, the classical output field, E_3, is

$$E_3 = rE_1 + tE_2 \tag{6.1}$$

and the output field, E_4, is

$$E_4 = tE_1 - rE_2 \tag{6.2}$$

where r and t are now real numbers and we have included the π phase shift in
Eq. (6.2) due to reflection from the side of the beam splitter with the "dot" as seen in
Fig. 6.1. We assume the same polarization for all fields, so we only need to consider
each field amplitude and phase (we can treat the electric field as a scalar, rather than
as a vector).

For simplicity, suppose the input beam at the top of Fig. 6.1 is absent, so $E_2 = 0$.
The output fields are then

$$E_3 = rE_1 \tag{6.3}$$

$$E_4 = tE_1 \tag{6.4}$$

and the corresponding intensities measured at the detectors, D_3 and D_4, are propor-
tional to the square modulus of the fields, $|E_3|^2$ and $|E_4|^2$:

$$I_3 \propto |E_3|^2 = |r|^2|E_1|^2 = R|E_1|^2 \tag{6.5}$$

$$I_4 \propto |E_4|^2 = |t|^2|E_1|^2 = T|E_1|^2 \tag{6.6}$$

where $R = |r|^2$ and $T = |t|^2$ are called the reflectance and transmittance, respectively. They refer to the fraction of light intensity, which is reflected or transmitted (rather than r and t, which refer to the fields). Adding Eqs. (6.5) and (6.6) gives

$$|E_3|^2 + |E_4|^2 = (R + T)|E_1|^2 \qquad (6.7)$$

Due to the conservation of energy,

$$|E_3|^2 + |E_4|^2 = |E_1|^2 \qquad (6.8)$$

That is, the input and output intensities must match. Comparing Eqs. (6.7) and (6.8), we must have

$$R + T = 1 \qquad (6.9)$$

which is just another way of stating the conservation of energy. For the general case with both inputs present ($E_1 \neq 0$, $E_2 \neq 0$), we must have

$$|E_3|^2 + |E_4|^2 = |E_1|^2 + |E_2|^2 \qquad (6.10)$$

This results in the general condition:

$$rr^* + tt^* = 1 \qquad (6.11)$$

$$rt^* - tr^* = 0 \qquad (6.12)$$

where r^* and t^* are the complex conjugate of r and t, respectively. Equation (6.11) is equivalent to Eq. (6.9).

Exercise 6.1 Derive Eqs. (6.11) and (6.12), referring to Fig. 6.1.

6.2 Quantum Beam Splitter

The quantum approach to the beam splitter, first introduced in 1987 [4–6], replaces the classical fields with the corresponding quantum operators, as shown in Fig. 6.2. The fields are

$$\widehat{E}_3 = r\widehat{E}_1 + t\widehat{E}_2 \qquad (6.13)$$

$$\widehat{E}_4 = t\widehat{E}_1 - r\widehat{E}_2 \qquad (6.14)$$

analogous to Eqs. (6.1) and (6.2).

Fig. 6.2 A beam splitter
with quantum operators for
the input and output fields

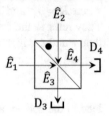

Fig. 6.3 A beam splitter
with annihilation operators
corresponding to the input
and output fields

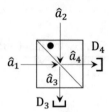

Since the field operators are related to the annihilation operators (Eq. (3.43)), the
annihilation operators are also given by the reflection and transmission coefficients
in the same manner as the fields, as shown in Fig. 6.3:

$$\hat{a}_3 = r\hat{a}_1 + t\hat{a}_2 \tag{6.15}$$

$$\hat{a}_4 = t\hat{a}_1 - r\hat{a}_2 \tag{6.16}$$

Taking the Hermitian conjugate of Eqs. (6.15) and (6.16) gives

$$\hat{a}_3^\dagger = r^*\hat{a}_1^\dagger + t^*\hat{a}_2^\dagger \tag{6.17}$$

$$\hat{a}_4^\dagger = t^*\hat{a}_1^\dagger - r^*\hat{a}_2^\dagger \tag{6.18}$$

or, since r and t are real numbers in our case:

$$\hat{a}_3^\dagger = r\hat{a}_1^\dagger + t\hat{a}_2^\dagger \tag{6.19}$$

$$\hat{a}_4^\dagger = t\hat{a}_1^\dagger - r\hat{a}_2^\dagger \tag{6.20}$$

Note that Eqs. (6.15) and (6.16) can be summarized in matrix form:

$$\begin{pmatrix} \hat{a}_3 \\ \hat{a}_4 \end{pmatrix} = \begin{pmatrix} r & t \\ t & -r \end{pmatrix} \begin{pmatrix} \hat{a}_1 \\ \hat{a}_2 \end{pmatrix} \tag{6.21}$$

The transformation matrix in Eq. (6.21) is unitary as required for any quantum
transformation.

> **Exercise 6.2** Using the commutator relations of the beam splitter input operators, show that the correct commutator relations are obtained for the output operators.

Let us find an expression for the input annihilation operators in terms of the output operators. Starting with Eq. (6.15) and multiplying both sides by r^* gives

$$r^*\hat{a}_3 = r^*r\hat{a}_1 + r^*t\hat{a}_2 \qquad (6.22)$$

Similarly, starting with Eq. (6.16) and multiplying both sides by t^* gives

$$t^*\hat{a}_4 = t^*t\hat{a}_1 - t^*r\hat{a}_2 \qquad (6.23)$$

Adding Eqs. (6.22) and (6.23) gives

$$(r^*r + t^*t)\hat{a}_1 + (r^*t - t^*r)\hat{a}_2 = r^*\hat{a}_3 + t^*\hat{a}_4 \qquad (6.24)$$

From Eqs. (6.11) and (6.12), Eq. (6.24) simplifies to

$$\hat{a}_1 = r^*\hat{a}_3 + t^*\hat{a}_4 \qquad (6.25)$$

Similarly, we can prove that

$$\hat{a}_2 = t^*\hat{a}_3 - r^*\hat{a}_4 \qquad (6.26)$$

Taking the Hermitian conjugate of Eqs. (6.25) and (6.26) gives

$$\hat{a}_1^\dagger = r\hat{a}_3^\dagger + t\hat{a}_4^\dagger \qquad (6.27)$$

$$\hat{a}_2^\dagger = t\hat{a}_3^\dagger - r\hat{a}_4^\dagger \qquad (6.28)$$

These relations will be useful in our upcoming derivations.

> **Exercise 6.3** Derive Eq. (6.26).

6.3   Input/Output Transformation

Suppose we have some input state to the beam splitter, $|\psi_{\text{in}}\rangle$. The output state can be obtained by applying a transformation matrix, similar to Eq. (6.21):

$$|\psi_{\text{out}}\rangle = U|\psi_{\text{in}}\rangle \tag{6.29}$$

The expectation value $\langle O_{\text{out}}\rangle$, associated with some output operator of the beam splitter, $\widehat{O}_{\text{out}}$, is

$$\langle O_{\text{out}}\rangle = \langle \psi_{\text{out}}|\widehat{O}_{\text{out}}|\psi_{\text{out}}\rangle \tag{6.30}$$

Using Eq. (6.29):

$$\langle O_{\text{out}}\rangle = \langle \psi_{\text{in}}|U^{\dagger}\widehat{O}_{\text{out}}U|\psi_{\text{in}}\rangle \tag{6.31}$$

The transformation $U^{\dagger}\widehat{O}_{\text{out}}U$ can be interpreted as the operator $\widehat{O}_{\text{out}}$ (expressed in terms of the output modes 3 and 4) transformed into an input operator (expressed in terms of the input modes 1 and 2):

$$\widehat{O}_{\text{in}} = U^{\dagger}\widehat{O}_{\text{out}}U \tag{6.32}$$

Thus,

$$\langle O_{\text{out}}\rangle = \langle \psi_{\text{in}}|\widehat{O}_{\text{in}}|\psi_{\text{in}}\rangle \tag{6.33}$$

In summary, $\langle O_{\text{out}}\rangle$ can be determined using either the input space or the output space:

$$\langle O_{\text{out}}\rangle = \langle \psi_{\text{out}}|\widehat{O}_{\text{out}}|\psi_{\text{out}}\rangle = \langle \psi_{\text{in}}|\widehat{O}_{\text{in}}|\psi_{\text{in}}\rangle \tag{6.34}$$

This idea will become more clear in the next section where it will be used to simplify the analysis of the beam splitter.

6.4   Single Photon on a Beam Splitter

Consider the beam splitter in Fig. 6.4 with a single photon input state, $|1\rangle_1$, on port 1 and nothing on port 2. "Nothing" is represented by the vacuum state, $|0\rangle_2$, on port 2. The combined input state is denoted as $|\psi_{\text{in}}\rangle = |1\rangle_1|0\rangle_2$. Single photons are launched into the beam splitter and the detection events at D_3 and D_4 are counted. After repeating the experiment many times, the probability of detection at D_3 and D_4 is determined. For *each* photon launched into the beam splitter, the probability of detection at D_3 is

Fig. 6.4 Single photon on input port 1

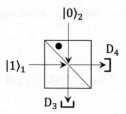

$$P_3 = \frac{\langle n_3 \rangle}{\langle n_1 \rangle} = \frac{\langle n_3 \rangle}{1} = \langle \psi_{\text{out}} | \widehat{N}_3 | \psi_{\text{out}} \rangle \tag{6.35}$$

where $\widehat{N}_3$ is the number operator for D_3 and $|\psi_{\text{out}}\rangle$ is the output state at D_3. $\langle n_1 \rangle$ represents the single photon input to port 1, that is, $\langle n_1 \rangle = \left\langle \psi_{\text{in}} \left| \widehat{a}_1^\dagger \widehat{a}_1 \right| \psi_{\text{in}} \right\rangle = {}_2\langle 0|_1 \langle 1 | \widehat{a}_1^\dagger \widehat{a}_1 | 1 \rangle_1 | 0 \rangle_2 = 1$. According to the definition of the number operator, we have

$$P_3 = \langle \psi_{\text{out}} | \widehat{a}_3^\dagger \widehat{a}_3 | \psi_{\text{out}} \rangle \tag{6.36}$$

We can convert the output space of Eq. (6.36) into the input space using Eq. (6.34). We replace $|\psi_{\text{out}}\rangle$ with $|\psi_{\text{in}}\rangle$ and the output operators, $\widehat{a}_3^\dagger$ and $\widehat{a}_3$, are expressed in terms of the input operators using Eqs. (6.15) and (6.17):

$$P_3 = \langle \psi_{\text{in}} | (r^* \widehat{a}_1^\dagger + t^* \widehat{a}_2^\dagger)(r \widehat{a}_1 + t \widehat{a}_2) | \psi_{\text{in}} \rangle \tag{6.37}$$

Using $|\psi_{\text{in}}\rangle = |1\rangle_1 |0\rangle_2$ gives

$$P_3 = {}_2\langle 0|_1 \langle 1 | (r^* r \widehat{a}_1^\dagger \widehat{a}_1 + r^* t \widehat{a}_1^\dagger \widehat{a}_2 + t^* r \widehat{a}_2^\dagger \widehat{a}_1 + t^* t \widehat{a}_2^\dagger \widehat{a}_2) | 1 \rangle_1 | 0 \rangle_2 \tag{6.38}$$

Using the properties of the creation and annihilation operators on the states $|0\rangle$ and $|1\rangle$, only the first term in Eq. (6.38) survives, giving:

$$P_3 = {}_2\langle 0|_1 \langle 1 | r^* r | 1 \rangle_1 | 0 \rangle_2 = r^* r \, {}_2\langle 0|0 \rangle_2 \, {}_1\langle 1|1 \rangle_1 \tag{6.39}$$

In Eq. (6.39), ${}_2\langle 0|0 \rangle_2 = 1$ and ${}_1\langle 1|1 \rangle_1 = 1$ since the states are normalized, giving

$$P_3 = r^* r = |r|^2 = R \tag{6.40}$$

Here, R is the probability of a single photon from port 1 being detected at D_3. Note that R is the reflectance from the beam splitter. Equation (6.40) is the same as the classical result for the fraction R of light intensity reflected by the beam splitter from port 1 to D_3.

Similarly, we can calculate the probability of detection at D_4:

$$P_4 = \langle \psi_{\text{out}} | \hat{a}_4^\dagger \hat{a}_4 | \psi_{\text{out}} \rangle \tag{6.41}$$

Using Eqs. (6.16) and (6.18) gives

$$P_4 = \langle \psi_{\text{in}} | (t^* \hat{a}_1^\dagger - r^* \hat{a}_2^\dagger)(t \hat{a}_1 - r \hat{a}_2) | \psi_{\text{in}} \rangle \tag{6.42}$$

$$= {}_2\langle 0 |_1 \langle 1 | (t^* t \hat{a}_1^\dagger \hat{a}_1 - t^* r \hat{a}_1^\dagger \hat{a}_2 - r^* t \hat{a}_2^\dagger \hat{a}_1 + r^* r \hat{a}_2^\dagger \hat{a}_2) | 1 \rangle_1 | 0 \rangle_2 \tag{6.43}$$

$$= {}_2\langle 0 |_1 \langle 1 | (t^* t) | 1 \rangle_1 | 0 \rangle_2 \tag{6.44}$$

$$= t^* t = |t|^2 = T \tag{6.45}$$

Here, T is the probability of a single photon from port 1 being detected at D_4. Note that T is the transmittance through the beam splitter. Equation (6.45) is the same as the classical result for the fraction T of light intensity transmitted by the beam splitter from port 1 to D_4. Note that the total probability for detection at either detector adds to unity as required, since $P_3 + P_4 = R + T = 1$.

6.5 Coincident Measurements

Let us calculate the probability, P_{34}, of simultaneous detection at D_3 and D_4. Simultaneous detection events are called coincidence or correlation measurements. Classically, we would expect a fraction R of a classical light intensity reflected to D_3 and a fraction T transmitted to D_4, allowing simultaneous detection. Classically, we expect the intensity to be proportional to $(E_3)^2 = R(E_1)^2$ at detector D_3 and $(E_4)^2 = T(E_1)^2$ at detector D_4, giving the probability of double detection:

$$\text{Classical}: P_{34} = \frac{R(E_1)^2 T(E_1)^2}{(E_1)^2 (E_1)^2} = RT \tag{6.46}$$

Let us calculate the probability of a coincident detection, P_{34}, for a single photon input on the beam splitter, as illustrated in Fig. 6.5. The simultaneous measurement is described by the operator $\hat{a}_4 \hat{a}_3$:

$$P_{34} = \langle \psi_{\text{out}} | (\hat{a}_4 \hat{a}_3)^\dagger (\hat{a}_4 \hat{a}_3) | \psi_{\text{out}} \rangle = \langle \psi_{\text{out}} | \hat{a}_3^\dagger \hat{a}_4^\dagger \hat{a}_4 \hat{a}_3 | \psi_{\text{out}} \rangle \tag{6.47}$$

Using Eqs. (6.15) to (6.18) gives

Fig. 6.5 Correlation or coincidence measurement (simultaneous detection at D_3 and D_4) with single photon input on port 1

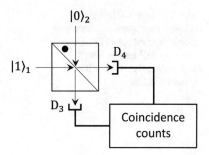

$$P_{34} = {}_2\langle 0|\,{}_1\langle 1|\,(r^*\hat{a}_1^\dagger + t^*\hat{a}_2^\dagger)(t^*\hat{a}_1^\dagger - r^*\hat{a}_2^\dagger)(t\hat{a}_1 - r\hat{a}_2)(r\hat{a}_1 + t\hat{a}_2)|1\rangle_1|0\rangle_2 \quad (6.48)$$

Evaluating all the terms of Eq. (6.48) gives

$$P_{34} = 0 \tag{6.49}$$

Quantum mechanically, double detections are not possible for a single photon, which is very different than the classical result of Eq. (6.46). The single photon is detected at D_3 with probability R, or at D_4 with probability T, but never both simultaneously. Here, we have a nonclassical correlation. The absence of double detections must be the case if the concept of "single photon" is to make any sense at all. You can only detect a single photon once, either at D_3 or D_4.

Exercise 6.4 Evaluate Eq. (6.48), verifying that $P_{34} = 0$.

The single photon beam splitter could be used as a random number generator. With a 50:50 beam splitter ($R = T = 0.5$), we have a probability $P_3 = P_4 = 0.5$ that a single photon is detected at either D_3 or D_4. A single photon is launched into the beam splitter, and a 0 bit is assigned for detection at D_3, while a 1 bit is assigned for detection at D_4. After launching many single photons, one at a time, into the beam splitter, a random sequence of bits is generated, 00110101110... The random sequence of bits can be used to generate a random number.

6.6   Second-Order Correlation Function

The correlations described in the previous section are usually described by a second-order correlation function, $g^{(2)}(\tau)$, introduced in 1963 by Roy Glauber (Fig. 6.6), a pioneer of quantum optics [7]. The 2005 Nobel Prize in Physics was divided, one half awarded to Roy J. Glauber "for his contribution to the quantum theory of optical coherence," the other half jointly to John L. Hall and Theodor W. Hänsch "for their contributions to the development of laser-based precision spectroscopy, including the optical frequency comb technique."

Fig. 6.6 Roy J. Glauber.
(Credit: Photograph by Jane
Reed, Harvard News Office,
courtesy AIP Emilio Segrè
Visual Archives, Gift of Roy
Glauber, 2006)

First, we look at the classical definition of the second-order correlation function, which is given by

$$g_{\text{classical}}^{(2)}(\tau) = \frac{\langle I(t)I(t+\tau)\rangle}{\langle I(t)\rangle^2} = \frac{\langle E^*(t)E(t)E^*(t+\tau)E(t+\tau)\rangle}{\langle E^*(t)E(t)\rangle^2} \tag{6.50}$$

where $I \propto |E|^2 = E^*E$. The brackets, $\langle\rangle$, indicate an average to account for intensity fluctuations during the measurement time. $g_{\text{classical}}^{(2)}(\tau)$ describes the correlation between two temporally separated intensity signals with time difference τ from one source. If $\tau = 0$, $g_{\text{classical}}^{(2)}(0)$ is especially interesting, because it gives the probability of simultaneous detection events at two detectors, normalized to the probability of individual detection events at either detector. The " 0 " means no time delay between the two simultaneous detections.

Suppose the input to the beam splitter is treated as a classical source of light. For classical light, we have

$$g_{\text{classical}}^{(2)}(0) = \frac{\langle RE_1^2 TE_1^2\rangle}{\left(R\langle E_1\rangle^2\right)\left(T\langle E_1\rangle^2\right)} \tag{6.51}$$

R and T cancel out, and since E_1^2 is proportional to the light intensity, we get

$$g_{\text{classical}}^{(2)}(0) = \frac{\langle I^2\rangle}{\langle I\rangle^2} \tag{6.52}$$

Next, we can use the Cauchy–Schwarz inequality, which states

$$\langle I^2 \rangle \geq \langle I \rangle^2 \tag{6.53}$$

for any positive random variable. Thus,

$$g^{(2)}_{classical}(0) \geq 1 \tag{6.54}$$

We can rewrite Eq. (6.50) for quantum light by replacing the electric field with its corresponding operator in the normal order:

$$g^{(2)}(\tau) = \frac{\langle \widehat{E}^-(t)\widehat{E}^-(t+\tau)\widehat{E}^+(t+\tau)\widehat{E}^+(t) \rangle}{\langle \widehat{E}^-(t)\widehat{E}^+(t) \rangle^2} \tag{6.55}$$

where $\widehat{E}^-$ and $\widehat{E}^+$ are the negative and positive frequency components, respectively, of the field operator introduced in Chap. 5. The correlation function, $g^{(2)}(0)$, for quantum light becomes

$$g^{(2)}(0) = \frac{\langle \widehat{a}^\dagger \widehat{a}^\dagger \widehat{a}\widehat{a} \rangle}{\langle \widehat{a}^\dagger \widehat{a} \rangle^2} \tag{6.56}$$

Using the commutation relation, $\widehat{a}^\dagger \widehat{a} = \widehat{a}\widehat{a}^\dagger - 1$, gives

$$g^{(2)}(0) = \frac{\langle \widehat{a}^\dagger (\widehat{a}\widehat{a}^\dagger - 1)\widehat{a} \rangle}{\langle \widehat{a}^\dagger \widehat{a} \rangle^2} \tag{6.57}$$

$$= \frac{\langle \widehat{a}^\dagger \widehat{a}\widehat{a}^\dagger \widehat{a} - \widehat{a}^\dagger \widehat{a} \rangle}{\langle \widehat{a}^\dagger \widehat{a} \rangle^2} \tag{6.58}$$

Recognizing $\widehat{a}^\dagger \widehat{a}$ as the number operator, we can write Eq. (6.58) in various equivalent forms:

$$g^{(2)}(0) = \frac{\langle n^2 - n \rangle}{\langle n \rangle^2} = \frac{\langle n(n-1) \rangle}{\langle n \rangle^2} = \frac{\langle n^2 \rangle - \langle n \rangle}{\langle n \rangle^2} \tag{6.59}$$

Using the variance, $(\Delta n)^2 = \langle n^2 \rangle - \langle n \rangle^2$, gives

$$g^{(2)}(0) = \frac{(\Delta n)^2 + \langle n \rangle^2 - \langle n \rangle}{\langle n \rangle^2} \tag{6.60}$$

$$= 1 + \frac{(\Delta n)^2 - \langle n \rangle}{\langle n \rangle^2} \tag{6.61}$$

Let us examine the case for a Fock state $|n\rangle$ on a beam splitter. The probability of detection at D_3 and D_4 are:

$$P_3 = R \langle n | \hat{a}^\dagger \hat{a} | n \rangle = Rn \tag{6.62}$$

$$P_4 = T \langle n | \hat{a}^\dagger \hat{a} | n \rangle = Tn \tag{6.63}$$

and the product is

$$P_3 P_4 = RTn^2 \tag{6.64}$$

The probability of double detection is

$$P_{34} = RT \langle n | \hat{a}^\dagger \hat{a}^\dagger \hat{a} \hat{a} | n \rangle \tag{6.65}$$

$$= RT \langle n | \hat{a}^\dagger (\hat{a} \hat{a}^\dagger - 1) \hat{a} | n \rangle \tag{6.66}$$

$$= RT \langle n | (\hat{a}^\dagger \hat{a} \hat{a}^\dagger \hat{a} - \hat{a}^\dagger \hat{a}) | n \rangle \tag{6.67}$$

$$= RT (n^2 - n) \tag{6.68}$$

$$= RTn(n - 1) \tag{6.69}$$

Thus, for the Fock state, we get

$$\text{Fock state}: g^{(2)}(0) = \frac{P_{34}}{P_3 P_4} = \frac{n(n-1)}{n^2} = \frac{n-1}{n} < 1 \tag{6.70}$$

which we could have also obtained directly from Eq. (6.59). The result in Eq. (6.70) is very different from classical light where $g^{(2)}_{classical}(0) \geq 1$. In the quantum case, the detection of a photon destroys it and changes the state, leaving one photon less (thus, we get $n(n - 1)$ in the numerator of Eq. (6.70)). This reduces the probability of double detection, which is not considered in the classical expression. This quantum effect ($g^{(2)}(0) < 1$) is called "anticorrelation" or "antibunching". The first measurement of antibunching in Ref. [8] showed the nonclassical or quantum nature of light. If we set $n = 1$ for a single photon state, we obtain $g^{(2)}(0) = 0$ from Eq. (6.70), as expected. The detection of a single photon destroys it, leaving zero probability of double detection.

6.7 Entangled State

If we have no input to the beam splitter, then we expect the action of the beam splitter to give

$$
\underbrace{|0\rangle_1|0\rangle_2}_{\text{input}} \xrightarrow[\text{splitter}]{\text{beam}} \underbrace{|0\rangle_3|0\rangle_4}_{\text{output}} \tag{6.71}
$$

The single photon input state in Fig. 6.4 can be expressed as

$$
|\psi_{\text{in}}\rangle = |1\rangle_1|0\rangle_2 = \hat{a}_1^\dagger|0\rangle_1|0\rangle_2 \tag{6.72}
$$

According to Eqs. (6.34) and (6.71), Eq. (6.72) can be expressed in terms of the output space using Eq. (6.27):

$$
\hat{a}_1^\dagger|0\rangle_1|0\rangle_2 \xrightarrow[\text{splitter}]{\text{beam}} \left(r\hat{a}_3^\dagger + t\hat{a}_4^\dagger\right)|0\rangle_3|0\rangle_4 \tag{6.73}
$$

$$
= r|1\rangle_3|0\rangle_4 + t|0\rangle_3|1\rangle_4 \tag{6.74}
$$

The output in Eq. (6.74) is called an entangled state of a photon in the D_3 path and the D_4 path. An entangled state is a state, which cannot be separated or factored into individual product states; that is, $|\psi\rangle \neq |\psi_3\rangle_3|\psi_4\rangle_4$. Equation (6.74) cannot be factored into the product of two individual states (try it). If you are not familiar with entanglement, do not worry. We will cover this topic in more detail in Chap. 8.

Equation (6.74) tells us that the single photon input on port 1 results in a superposition of the single photon in mode 3 with zero photons in mode 4, and vice versa. The probability amplitude for a single photon along the D_3 path is given by r, that is, the coefficient of $|1\rangle_3|0\rangle_4$ in Eq. (6.74). The corresponding probability is the modulus squared of the probability amplitude, $|r|^2 = R$, the same as Eq. (6.40). The probability amplitude for a single photon along the D_4 path is given by t, that is, the coefficient of $|0\rangle_3|1\rangle_4$ in Eq. (6.74). The corresponding probability is the modulus squared of the probability amplitude, $|t|^2 = T$, the same as Eq. (6.45). As seen in Eq. (6.74), the probability of joint detection at D_3 and D_4, represented by the state $|1\rangle_3|1\rangle_4$, is zero.

6.8   Hanbury Brown–Twiss Experiment

The concept of coincidence measurements, illustrated in Fig. 6.5, was first proposed for radio astronomy by Robert Hanbury Brown and Richard Twiss in 1954 [9], and later extended to optical signals [10]. Known as the Hanbury Brown–Twiss experiment, it is now a standard method to characterize light sources, including whether an emitter is a good source of single photons. For a single photon input to the beam splitter, we expect $g^{(2)}(0) = 0$, which is very different than the classical result of $g^{(2)}(0) \geq 1$ from Eq. (6.54). $g^{(2)}(0)$ below 10^{-3} has been measured for certain single photon sources [11].

References

1. F.L. Pedrotti, L.M. Pedrotti and L.S. Pedrotti, *Introduction to optics* (3rd Edition, Pearson, 2006).
2. M. W. Hamilton, *Phase shifts in multilayer dielectric beam splitters*, Amer. J. Phys. 68 (2000) 186.
3. F. Hénault, *Quantum physics and the beam splitter mystery*, Proc. SPIE 9570, The Nature of Light: What are Photons? VI, 95700Q (10 September 2015); https://doi.org/10.1117/12.2186291
4. S. Prasad, M.O. Scully and W. Martienssen, *A quantum description of the beam splitter*, Opt. Commun. 62 (1987) 139.
5. Z.Y. Ou C.K. Hong and L. Mandel, *Relation between input and output states for a beam splitter*, Opt. Commun. 63 (1987) 118.
6. H. Fearn and R. Loudon, *Quantum theory of the lossless beam splitter*, Opt. Commun. 64 (1987) 485.
7. R. J. Glauber, *Quantum theory of optical coherence*, Phys. Rev. 130 (1963) 2529.
8. H. J. Kimble, M. Dagenais, L. Mandel, *Photon antibunching in resonance fluorescence*, Phys. Rev. Lett. 39 (1977) 691.
9. R. Hanbury Brown and R.Q. Twiss, *A new type of interferometer for use in radio astronomy*, Philosophical Magazine 45 (1954) 663.
10. R. Hanbury Brown and R.Q. Twiss, *Correlation between photons in two coherent beams of light*, Nature 177 (1956) 27.
11. M.D. Eisaman, J. Fan, A. Migdall and S.V. Polyakov, *Single-photon sources and detectors*, Rev. Sci. Instrum. 82 (2011) 071101.

Chapter 7
Single Photon in an Interferometer

Another important device in quantum optics is the interferometer. Building on the results of the previous chapter for a beam splitter, the classical and quantum optics treatment of the Mach-Zehnder interferometer is introduced. The case of a single photon in an interferometer is treated, which introduces the concept of wave-particle duality.

7.1 Classical Light Interference

The Mach-Zehnder (MZ) interferometer, illustrated for classical light in Fig. 7.1, superposes the light field on a detector from two possible paths. Note that the two beam splitters in Fig. 7.1 are reversed. The input field E_1 along path 1 to detector D_3 undergoes two transmissions (t^2) and a phase shift of e^{ikz_1} associated with the distance z_1. The input field E_1 along path 2 to detector D_3 undergoes two reflections (r and $-r$) and a phase shift of e^{ikz_2} associated with the distance z_2. Therefore, the field at detector D_3 is a result of the superposition of the field along the two paths:

$$E_3 = \left(t^2 e^{ikz_1} - r^2 e^{ikz_2}\right)E_1 \qquad (7.1)$$

We assume r and t are real numbers, so $r^* = r$ and $t^* = t$. The light intensity (or detection probability) at D_3 is proportional to the square of the field, $I_3 \propto (E_3)^2 = E_3 E_3^*$. From Eq. (7.1), we get

$$I_3 = \left(t^2 e^{ikz_1} - r^2 e^{ikz_2}\right)\left(t^2 e^{-ikz_1} - r^2 e^{-ikz_2}\right)I_1 \qquad (7.2)$$

where $I_1 \propto (E_1)^2 = E_1 E_1^*$. Evaluating Eq. (7.2) gives

© The Author(s), under exclusive license to Springer Nature Switzerland AG 2022
R. LaPierre, *Getting Started in Quantum Optics*, Undergraduate Texts in Physics,
https://doi.org/10.1007/978-3-031-12432-7_7

Fig. 7.1 Mach-Zehnder
interferometer with classical
light

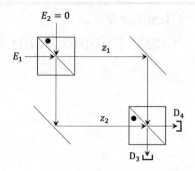

$$I_3 = \left[R^2 + T^2 - RT\left(e^{ik(z_1-z_2)} + e^{-ik(z_1-z_2)} \right) \right] I_1 \tag{7.3}$$

$$= \left[R^2 + T^2 - 2RT \cos\left(k\Delta z \right) \right] I_1 \tag{7.4}$$

where $\Delta z = z_1 - z_2$ is the path length difference between path 1 and path 2, and $k\Delta z$ is the corresponding phase difference.

Similarly, the field at detector D_4 is

$$E_4 = \left(rte^{ikz_1} + tre^{ikz_2} \right) E_1 \tag{7.5}$$

and the intensity is

$$I_4 = \left(rte^{ikz_1} + tre^{ikz_2} \right)\left(rte^{-ikz_1} + tre^{-ikz_2} \right) I_1 \tag{7.6}$$

$$= \left[2RT + RT\left(e^{ik(z_1-z_2)} + e^{-ik(z_1-z_2)} \right) \right] I_1 \tag{7.7}$$

$$= \left[2RT + 2RT \cos\left(k\Delta z \right) \right] I_1 \tag{7.8}$$

Adding Eqs. (7.4) and (7.8), we get

$$I_3 + I_4 = \left(R^2 + T^2 + 2RT \right) I_1 \tag{7.9}$$

$$= (R + T)^2 I_1 \tag{7.10}$$

$$= I_1 \tag{7.11}$$

since $R + T = 1$. The total output light intensity at D_3 and D_4 is equal to the input light intensity, as expected from the conservation of energy.

For 50:50 beam splitters ($R = T = 0.5$), we get

$$I_3 = \frac{1}{2}[1 - \cos\left(k\Delta z \right)] I_1 = I_1 \sin^2\left(\frac{k\Delta z}{2} \right) \tag{7.12}$$

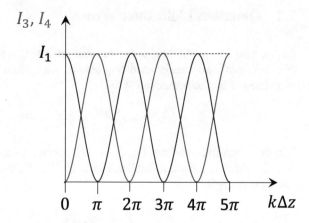

Fig. 7.2 Light output intensity at D_3 (red) and D_4 (blue) versus phase difference, $k\Delta z$

$$I_4 = \frac{1}{2}[1 + \cos(k\Delta z)]\, I_1 = I_1 \cos^2\left(\frac{k\Delta z}{2}\right) \tag{7.13}$$

Let us suppose that we change the path length difference, Δz, between path 1 and 2 (for example, by moving one of the mirrors) and we measure the intensities at D_3 and D_4 as a function of the phase difference, $k\Delta z$. The result from Eqs. (7.12) and (7.13) is plotted in Fig. 7.2. The oscillating intensities are due to alternating constructive and destructive interference of the two light fields along path 1 and 2 as the path length difference, Δz, changes. In particular, if $k\Delta z = 0$ or any integer multiple of 2π, then all of the input light appears at port 4 and none at port 3. We call port 3 the "dark port". A small change in the path length difference, Δz, will appear as some light on the dark port. In this way, we can detect some path length difference, Δz.

Let us calculate the probability of double detection, that is, simultaneous detection at both D_3 and D_4. Classically, for 50:50 beam splitters in the interferometer, we get

$$P_{34} = \frac{\left[I_1 \sin^2\left(\frac{k\Delta z}{2}\right)\right]\left[I_1 \cos^2\left(\frac{k\Delta z}{2}\right)\right]}{(I_1)^2} \tag{7.14}$$

$$= \sin^2\left(\frac{k\Delta z}{2}\right)\cos^2\left(\frac{k\Delta z}{2}\right) \tag{7.15}$$

$$= \frac{1}{4}\sin^2(k\Delta z) \tag{7.16}$$

7.2   Quantum Light Interference

Let us now derive the output for the case of a single photon input, as shown in Fig. 7.3. Here, we clearly need a quantum description. The annihilation operator associated with photodetection at D_3 is

$$\hat{a}_3 = \left(t^2 e^{ikz_1} - r^2 e^{ikz_2}\right)\hat{a}_1 + \left(-tre^{ikz_1} - rte^{ikz_2}\right)\hat{a}_2 \qquad (7.17)$$

The first term follows from Eq. (7.1). The second term derives from the input on port 2, which is vacuum. Similar to Eq. (6.35), the probability of single photon detection at D_3 is

$$P_3 = \langle\psi_{\text{out}}|\hat{N}_3|\psi_{\text{out}}\rangle \qquad (7.18)$$

Using Eq. (7.17), we can express Eq. (7.18) in terms of the input space. Assuming r and t are real, we get

$$P_3 = {}_2\langle 0|_1\langle 1|\left[\left(t^2 e^{-ikz_1} - r^2 e^{-ikz_2}\right)\left(t^2 e^{ikz_1} - r^2 e^{ikz_2}\right)\hat{a}_1^\dagger\hat{a}_1\right]|1\rangle_1|0\rangle_2 \qquad (7.19)$$

where all terms related to $\hat{a}_2$ are omitted, because they result in zero when applied to the vacuum input, $|0\rangle_2$, on port 2. Evaluating Eq. (7.19) gives

$$P_3 = \left[R^2 + T^2 - 2RT\cos\left(k\Delta z\right)\right] \qquad (7.20)$$

which is the same as the classical result.

Similarly, the annihilation operator associated with photodetection at D_4 is

$$\hat{a}_4 = \left(rte^{ikz_1} + tre^{ikz_2}\right)\hat{a}_1 + \left(-r^2 e^{ikz_1} + t^2 e^{ikz_2}\right)\hat{a}_2 \qquad (7.21)$$

The probability of single photon detection at D_4 is

Fig. 7.3 Mach-Zehnder interferometer with single photon input

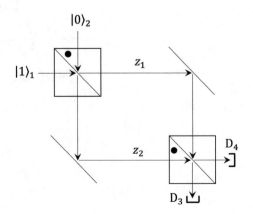

$$P_4 = \langle \psi_{\text{out}} | \widehat{N}_4 | \psi_{\text{out}} \rangle \tag{7.22}$$

or, in terms of the input space:

$$P_4 = {}_2\langle 0 |_1 \langle 1 | \left[\left(rte^{-ikz_1} + tre^{-ikz_2} \right) \left(rte^{ikz_1} + tre^{ikz_2} \right) \hat{a}_1^\dagger \hat{a}_1 \right] | 1 \rangle_1 | 0 \rangle_2 \tag{7.23}$$

$$= 2RT + 2RT \cos \left(k\Delta z \right) \tag{7.24}$$

which is the same as the classical result.

Exercise 7.1 Derive P_3 and P_4 if the two beam splitters in Fig. 7.3 are identically oriented with the "dot" on top. How does it compare to the results obtained above?

The probability of double (simultaneous) detection at D_3 and D_4 is

$$P_{34} = \langle \psi_{\text{out}} | \hat{a}_3^\dagger \hat{a}_4^\dagger \hat{a}_4 \hat{a}_3 | \psi_{\text{out}} \rangle \tag{7.25}$$

Evaluating Eq. (7.25), using Eqs. (7.17) and (7.21), gives

$$P_{34} = 0 \tag{7.26}$$

which is different from the classical case (but identical to the beam splitter in Chap. 6). Simultaneous detection events at D_3 and D_4 are possible with classical light, but not with single photons.

Exercise 7.2 Derive Eq. (7.26).

Suppose a single photon is input to the MZ interferometer for a given path length difference Δz, resulting in a detection event at either D_3 or D_4. The experiment is repeated many times, and the number of counts at D_3 and D_4 is tallied for a given Δz. Next, the experiment is repeated for different path length differences, Δz. The number of counts at D_3 and D_4 versus the phase, $k\Delta z$, can then be plotted. This experiment has been done, using an atomic cascade as a heralded source of single photons, with the results shown in Fig. 7.4 [1]. The quantum result in Fig. 7.4 is identical to the classical result in Fig. 7.2. An interference pattern occurs after counting many single photon detection events, although only a single photon traverses the interferometer at a time and only a single photon is ever detected at D_3 or D_4!

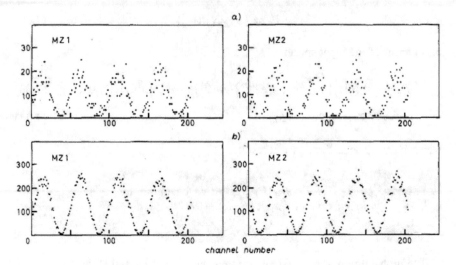

Fig. 7.4 Number of counts in outputs D_3 (labeled as MZ1) and D_4 (MZ2) as a function of the path difference Δz (one channel corresponds to a $\lambda/50$ variation of Δz). (**a**) 1 s counting time per channel. (**b**) 15 s counting time per channel. (Reproduced with permission from Grangier et al. [1])

7.3   Wave-Particle Duality

The fact that a single photon is detected at either D_3 or D_4, and there are no double detections in beam splitter or interferometer experiments, indicates that the photon exhibits particle behavior. On the other hand, after counting many single photon detection events, an interference pattern occurs that is indicative of wave behavior. This happens because the photon probability amplitude is put into a superposition of both paths, resulting in interference. Photodetection collapses the state to either D_3 or D_4. This is a manifestation of "wave-particle duality".

Exercise 7.3 In the 1970s, the physicist John Archibald Wheeler proposed a thought experiment, called the delayed choice experiment, which further demonstrates wave-particle duality. Describe the delayed choice experiment and its implications for the description of light.

Reference

1. P. Grangier, G. Roger and A. Aspect, *Experimental evidence for a photon anticorrelation effect on a beam splitter: A new light on single-photon interferences*, Europhys. Lett. 1 (1986) 173.

Chapter 8
Entanglement

There exist multiphoton states that cannot be expressed as a product of individual photon states. These states are called entangled states. Einstein used hidden variables in an attempt to explain the probabilities of quantum mechanics. John Bell proposed a test, using entangled states, showing that quantum mechanics cannot be explained by hidden variables.

8.1 Entangled States

In Chap. 6 (Sect. 6.7), we saw that the beam splitter output was an entangled state. Let us examine a two-photon state, which can be written in general as a product of two superpositions:

$$|\psi\rangle = |\psi\rangle_1 |\psi\rangle_2 = \left(\alpha_1 |0\rangle_1 + \beta_1 |1\rangle_1\right)\left(\alpha_2 |0\rangle_2 + \beta_2 |1\rangle_2\right) \tag{8.1}$$

where the subscript on each ket indicates the photon number (photon 1 or photon 2), and $|0\rangle$ and $|1\rangle$ represent the two possible orthogonal states of each photon—for example, horizontal and vertical polarization, or two different paths in a beam splitter or interferometer. Expanding this two-photon state gives

$$|\psi\rangle = \alpha_1\alpha_2 |0\rangle_1 |0\rangle_2 + \alpha_1\beta_2 |0\rangle_1 |1\rangle_2 + \beta_1\alpha_2 |1\rangle_1 |0\rangle_2 + \beta_1\beta_2 |1\rangle_1 |1\rangle_2 \tag{8.2}$$

Suppose you are given a composite state of two photons (e.g., Eq. (8.2)) and asked for the state of the individual photons. To answer this question, you would work backward to factor the state, obtaining Eq. (8.1). This is known as a separable state.

© The Author(s), under exclusive license to Springer Nature Switzerland AG 2022
R. LaPierre, *Getting Started in Quantum Optics*, Undergraduate Texts in Physics,
https://doi.org/10.1007/978-3-031-12432-7_8

There are some states for which this factoring is impossible; that is, you cannot write the composite state as a product of the individual states:

$$|\psi\rangle \neq |\psi\rangle_1 |\psi\rangle_2 \tag{8.3}$$

These are known as entangled states. In entangled states, you cannot talk about the state of the photons individually—they are somehow intertwined. Note that a general two-photon state, $|\psi\rangle = \alpha_{00}|0\rangle_1|0\rangle_2 + \alpha_{01}|0\rangle_1|1\rangle_2 + \alpha_{10}|1\rangle_1|0\rangle_2 + \alpha_{11}|1\rangle_1|1\rangle_2$, is usually entangled rather than separable – entanglement is normal in quantum mechanics!

There are four entangled two-photon states that are commonly encountered, known as the Bell states (we have dropped the particle subscripts):

$$|\Phi^+\rangle = \frac{1}{\sqrt{2}}(|00\rangle + |11\rangle) \tag{8.4}$$

$$|\Psi^+\rangle = \frac{1}{\sqrt{2}}(|01\rangle + |10\rangle) \tag{8.5}$$

$$|\Phi^-\rangle = \frac{1}{\sqrt{2}}(|00\rangle - |11\rangle) \tag{8.6}$$

$$|\Psi^-\rangle = \frac{1}{\sqrt{2}}(|01\rangle - |10\rangle) \tag{8.7}$$

The states in Eqs. (8.4), (8.5), (8.6) and (8.7) cannot be factored into the product of two individual photon states (try it) according to the definition of entanglement.

Suppose we prepare the entangled state $|\Psi^-\rangle$ (Eq. (8.7)) between two photons and then separate them by large distances (note that entangled states can be prepared by the various processes discussed in Chap. 5 such as atomic cascades, biexciton recombination in QDs, or spontaneous parametric down-conversion). Let us suppose that $|0\rangle$ represents vertical polarization and $|1\rangle$ represents horizontal polarization. After separating the photons, an individual (let us call her Alice) could perform a polarization measurement on her photon and another individual (let us call him Bob) could perform a subsequent polarization measurement on his photon. If Alice measures vertical polarization ($|0\rangle$), Bob will measure horizontal polarization ($|1\rangle$), and vice versa. The measurements are correlated between Alice and Bob, and this occurs no matter the distance between them. There appears to be instantaneous action at a distance or "nonlocality" in entanglement, which Einstein called "spooky action at a distance". Could Alice and Bob use entanglement to communicate instantaneously across vast distances? The theory of special relativity states that information cannot travel faster than the speed of light. Does entanglement violate special relativity?

In fact, entanglement does not violate special relativity, because no information is being transmitted. Upon measurement, Alice will collapse the entangled state, $|\Psi^-\rangle = \frac{1}{\sqrt{2}}(|01\rangle - |10\rangle)$, to a separable state—either $|01\rangle$ with probability ½, or $|10\rangle$

with probability ½; that is, Alice obtains either vertical or horizontal polarization with 50% probability (i.e., random), and Bob obtains the opposite polarization. Alice cannot control which of these two states she obtains. Subsequent measurement by Bob will result in the opposite polarization state to Alice, but his measurement will likewise appear to him to be completely random—no information is sent. Special relativity remains intact.

8.2   EPR Paradox and Hidden Variables

In a famous 1935 paper [1], Albert Einstein, Boris Podolsky, and Nathan Rosen (known as EPR) sought to demonstrate by the "EPR paradox" that quantum mechanics was incomplete. EPR were concerned by the instantaneous action at a distance, or "nonlocality", implied by entanglement. Quantum mechanics also seems to violate "realism". "Realism" means that particles have definite properties that are independent of any measurement.

Suppose we toss a coin. In principle, it is possible to know whether it will land heads or tails if we keep track of a lot of information about the system (called "degrees of freedom"), such as the forces applied during the toss, the air currents, the height of the toss, etc. However, all these physical properties are impossible to calculate in practice, so the most we can do is ascribe a probability distribution for the toss outcome resulting in $P_{\text{heads}} = \frac{1}{2}$ and $P_{\text{tails}} = \frac{1}{2}$. This outcome occurs from averaging the many degrees of freedom that we do not have access to. This principle also forms the basis for statistical thermodynamics.

Einstein and many others believed that quantum mechanics was like this; that is, they proposed that the probabilities in quantum mechanics are deterministic (versus probabilistic) and have some underlying causes that are "hidden"; that is, that we cannot access (analogous to the unknown variables during the coin toss). These underlying causes were called "hidden variables". If we knew the hidden variables, we would be able to calculate a definite measurement outcome, rather than just probabilities.

Many quantum pioneers, exemplified by Einstein, believed in "local realism" where the state of particles is defined when they are created. However, the "hidden variables" only allow us to determine the probability of these states. Einstein famously said: "God does not play dice with the universe". Also, with regards to realism, Einstein said "Do you believe the moon exists only when you look at it?"

Others, exemplified by Bohr, believed in the possibility of superpositions and entanglement. They believed that no definitive statements about a physical system may be made until a measurement is made. Particle properties do not exist until we measure them. It turns out that Bohr was correct; but how do we prove it?

8.3   CHSH Inequality

In 1964, the physicist, John Bell, proposed a test for quantum mechanics by measuring the spin states along three different directions $(\hat{a}, \hat{b}, \hat{c})$ for many entangled pairs of electron spins [2]. In 1969, Clauser, Horne, Shimony, and Holt (CHSH) refined the Bell test to one that was more amenable to experiment [3]. Rather than using entangled pairs of electron spins, CHSH proposed a test using entangled pairs of photon polarizations. Suppose a polarization entangled state, $|\psi\rangle = \frac{1}{\sqrt{2}} \times (|VH\rangle - |HV\rangle)$, is prepared where H represents horizontal polarization and V represents vertical polarization. Suppose a represents the result of a measurement along two orthogonal polarization directions for Alice. The measurement results are assigned the values $+1$ and -1; that is, $a = \pm 1$. For example, for the state $|\psi\rangle$, a would correspond to a polarization measurement. If the polarization measurement yields horizontal polarization, then the value $a = +1$ is assigned to the measurement result. Alternatively, if the polarization measurement yields vertical polarization, then the value $a = -1$ is assigned to the measurement result. Similarly, $a' = \pm 1$ represents the measurement results made by Alice along a different set of orthogonal polarization directions (e.g., 45° and 135°). Similarly, Bob can measure along two sets of orthogonal polarizations with results $b = \pm 1$ (corresponding to polarization along 22.5° or 112.5°) and $b' = \pm 1$ (corresponding to polarization along 67.5° and 157.5°). If $a = \pm 1$ and $a' = \pm 1$, it follows that either $a + a' = 0$, in which case $a - a' = \pm 2$. Otherwise, $a - a' = 0$, in which case $a + a' = \pm 2$. Therefore, we define a quantity S:

$$S = (a' + a)b + (a' - a)b' = a'b + ab + a'b' - ab' = \pm 2 \qquad (8.8)$$

Quantities such as ab represent a coincidence or correlation measurement. For example, a and b are obtained in measurements made by Alice and Bob, respectively, for each entangled photon pair and the product ab is determined. After repeated measurements on many photon pairs, the average value of S will depend on the probability distribution of each coincidence measurement. For example, the average of ab, $\langle ab \rangle$, will be given by the probability distribution $(P_{++})(1) + (P_{--})(1) + (P_{+-})(-1) + (P_{-+})(-1)$ where, for example, P_{++} is the probability of both Alice and Bob obtaining a measurement value of $+1$. After many trials, we must obtain

$$|\langle S \rangle| = |\langle a'b \rangle + \langle ab \rangle + \langle a'b' \rangle - \langle ab' \rangle| \leq 2 \qquad (8.9)$$

where the brackets $\langle \rangle$ represents an average over many coincidence measurements. Eq. (8.9) is the CHSH inequality. Different versions of the CHSH inequality exist, depending on how Eq. (8.9) is expressed. The signs in Eq. (8.9) are not so important as long as one of them is different than the others.

Fig. 8.1 Measurement
directions corresponding to
a, a', b, and b' equal to
+1 for the CHSH inequality
(-1 corresponds to the
orthogonal directions)

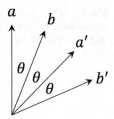

Equation (8.9) is a classical prediction. What does quantum mechanics predict? We can show for photons in the entangled state, $|\psi\rangle = \frac{1}{\sqrt{2}}(|HV\rangle - |VH\rangle)$, that $\langle ab \rangle = -\cos(2\theta)$ where θ is the angle between the two measurement directions (Exercise 8.1). For the angles mentioned earlier, and shown in Fig. 8.1, $\langle ab \rangle = -\cos(2 * 22.5°) = -\frac{1}{\sqrt{2}}$, $\langle a'b' \rangle = -\cos(2 * 22.5°) = -\frac{1}{\sqrt{2}}$, $\langle a'b \rangle = -\cos(2 * -22.5°) = -\frac{1}{\sqrt{2}}$, and $\langle ab' \rangle = -\cos(2 * 67.5°) = \frac{1}{\sqrt{2}}$, which gives $|\langle S \rangle| = 2\sqrt{2}$, that is, greater than the classical prediction of 2! It can be shown that the upper bound of $|\langle S \rangle|$ is $2\sqrt{2}$ in the case when a, b, a', and b' measurement directions are separated by successive 22.5° angles as shown in Fig. 8.1 [4].

Exercise 8.1 Show that for photons in the entangled state $|\psi\rangle = \frac{1}{\sqrt{2}} \times (|HV\rangle - |VH\rangle)$, we get $\langle ab \rangle = -\cos(2\theta)$ where θ is the angle between the measurement directions.

8.4   Testing the CHSH Inequality

The physicist, Alain Aspect (Fig. 8.2), famously performed an experiment in 1981 (Fig. 8.3), using polarization entangled photons produced by an atomic cascade, showing violation of the CHSH inequality because of quantum mechanics [5]. Polarizers were used to choose the polarization basis and split orthogonal polarization states to separate detectors. The path length between the source and the "Alice" detector is shorter than that for the "Bob" detector, so Alice performs her measurement before Bob. The experiment is repeated many times with entangled photon pairs, verifying a violation of the CHSH inequality. In the decades since Aspect's test, various "loopholes" in the test have been closed [7–9], confirming that quantum mechanics is not consistent with a hidden variables theory. Quantum mechanics violates local realism. Bohr was correct. However, the "spooky action at a distance" still seems mysterious and remains an inspiration to both physicists and philosophers.

Fig. 8.2 Alain Aspect
(Nobel Prize in Physics in
2022). (Credit: Wikimedia
Commons [6])

Fig. 8.3 Optical test of the
CHSH inequality

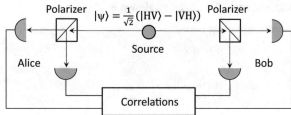

8.5   Quantum Key Distribution

Entangled states provide a means of secure communication called quantum key distribution (QKD). In 1991, Arthur Ekert proposed the E91 QKD protocol using entangled photons [10]. Alice and Bob receive one photon each of a polarization entangled pair from a source located either at Alice's or Bob's position, or somewhere else. For each photon they receive, Alice and Bob choose a polarization measurement basis and record the result of the measurement. Alice and Bob can perform a Bell test using their measurement results. If the Bell inequality is *not* violated, then an eavesdropper (called "Eve") must have destroyed the entanglement by eavesdropping. If the Bell inequality is violated, then there was no eavesdropper, and Alice and Bob can use their measurement results to generate a secret key to encrypt their messages. These implementations are referred to as entanglement-based protocols, Ekert protocols, or EPR protocols.

Entanglement provides a new resource that does not exist in classical physics. Further information on the applications of entanglement in quantum computing and QKD are available in Ref. [11]. In Chap. 18, applications of entanglement in metrology (quantum imaging) will be examined.

References

1. A. Einstein, B. Podolsky and N. Rosen, *Can quantum-mechanical description of physical reality be considered complete?*, Phys. Rev. 47 (1935) 777.
2. J.S. Bell, *On the Einstein Podolsky Rosen paradox*, Physics Physique Fizika 1 (1964) 195.
3. J.F. Clauser, M.A. Horne, A. Shimony and R.A. Holt, *Proposed experiment to test local hidden-variable theories*, Phys. Rev. Lett. 23 (1969) 880.
4. Lecture notes of John Preskill, Chapter 4, http://theory.caltech.edu/~preskill/ph219/index.html#lecture
5. A. Aspect, P. Grangier and G. Roger, *Experimental tests of realistic local theories via Bell's theorem*, Phys. Rev. Lett. 47 (1981) 460.
6. Attribution: The Royal Society. This file is licensed under the Creative Commons Attribution-Share Alike 3.0 Unported license (https://creativecommons.org/licenses/by-sa/3.0/deed.en). File: Alain-Aspect-ForMemRS.jpg. (2020, September 17). *Wikimedia Commons, the free media repository*. Retrieved 14:41, December 7, 2020 from https://commons.wikimedia.org/w/index.php?title=File:Alain-Aspect-ForMemRS.jpg&oldid=462669099
7. G. Weihs et al., *Violation of Bell's inequality under strict Einstein locality conditions*, Phys. Rev. Lett. 81 (1998) 5039.
8. C. Abellán et al., *Challenging local realism with human choices*, Nature 557 (2018) 212.
9. D. Rauch et al., *Cosmic Bell test using random measurement settings from high-redshift quasars*, Phys. Rev. Lett. 121 (2018) 080403.
10. A.K. Ekert, *Quantum cryptography based on Bell's theorem*, Phys. Rev. Lett. 67 (1991) 661.
11. R.R. LaPierre, *Introduction to quantum computing* (Springer, 2021).

Chapter 9
Multimode Quantized Radiation

In the previous chapters, we only considered single mode radiation, that is, radiation with a single frequency, ω, or wavevector, $\boldsymbol{k}$. In fact, single mode radiation is a "toy model" or approximation, since all real radiation is multimode. In reality, an infinite wave train of a single frequency doesn't exist. In this chapter, we introduce a more realistic description of light, called multimode light. The quantum optics treatment for the detection of a spontaneously emitted single photon wavepacket is presented as an example of multimode radiation.

9.1 Multimode Radiation

The general solution to Maxwell's equations in vacuum for classical electromagnetic waves is a superposition of fields:

$$E(r, t) = \sum_l \boldsymbol{\varepsilon}_l \varepsilon_l(t) e^{i k_l \cdot r} + c.c. \tag{9.1}$$

where l refers to the mode with polarization $\boldsymbol{\varepsilon}_l$, amplitude ε_l, wavevector $\boldsymbol{k}_l$, and frequency ω_l ($\omega_l = c k_l$). The amplitude of mode l is given by

$$\varepsilon_l(t) = \varepsilon_l(0) e^{-i \omega_l t} \tag{9.2}$$

9.2 Quantized Multimode Radiation

The canonical quantization of multimode radiation leads to the Hamiltonian:

© The Author(s), under exclusive license to Springer Nature Switzerland AG 2022
R. LaPierre, *Getting Started in Quantum Optics*, Undergraduate Texts in Physics,
https://doi.org/10.1007/978-3-031-12432-7_9

$$\hat{H} = \sum_l \hbar\omega_l \left(\hat{N}_l + \frac{1}{2}\right) = \sum_l \hbar\omega_l \left(\hat{a}_l^\dagger \hat{a}_l + \frac{1}{2}\right) \tag{9.3}$$

with energy eigenvalues

$$E = \sum_l \hbar\omega_l \left(n_l + \frac{1}{2}\right) \tag{9.4}$$

where $n_l = 0, 1, 2, \ldots$ is the number of photons for each of the modes $l = 1, 2, \ldots$
   Like Eq. (9.1), the most general solution for the quantized radiation field is a superposition of field operators with different frequencies (recall Eq. (3.43)):

$$\hat{E}(r) = \sum_l i\varepsilon_l \varepsilon_l^1 \left(\hat{a}_l e^{ik_l \cdot r} - \hat{a}_l^\dagger e^{-ik_l \cdot r}\right) \tag{9.5}$$

where ε_l^1 is the one photon amplitude of mode l with frequency ω_l:

$$\varepsilon_l^1 = \sqrt{\frac{\hbar\omega_l}{2\epsilon_o V}} \tag{9.6}$$

Each of the modes is decoupled and can be treated as independent quantum harmonic oscillators.
   The creation and annihilation operator for each mode l do not commute:

$$\left[\hat{a}_l, \hat{a}_l^\dagger\right] = 1 \tag{9.7}$$

while the operators for different modes, l and m, do commute:

$$\left[\hat{a}_l, \hat{a}_m^\dagger\right] = 0 \tag{9.8}$$

Equations (9.7) and (9.8) may be written succinctly as

$$\left[\hat{a}_l, \hat{a}_m^\dagger\right] = \delta_{lm} \tag{9.9}$$

The annihilation operator destroys one quantum of excitation (photon) of mode l:

$$\hat{a}_l |n_l\rangle = \sqrt{n_l} \, |n_l - 1\rangle \tag{9.10}$$

while the creation operator creates one photon of mode l:

$$\hat{a}_l^\dagger |n_l\rangle = \sqrt{n_l + 1} \, |n_l + 1\rangle \tag{9.11}$$

The Fock state of mode l can be generated from the vacuum state:

$$|n_l\rangle = \frac{\left(\widehat{a}_l^\dagger\right)^{n_l}}{\sqrt{n_l!}}|0_l\rangle \tag{9.12}$$

In general, a multimode Fock state can be written as

$$|\psi\rangle = |n_1\rangle \otimes |n_2\rangle \otimes \ldots = |n_1\rangle|n_2\rangle \ldots = |n_1, \ n_2, \ \ldots\rangle \tag{9.13}$$

where $\otimes$ denotes the tensor product. Alternatively, the tensor product can be written more succinctly as $|n_1\rangle|n_2\rangle \ldots$ or $|n_1, n_2, \ldots\rangle$ as shown in Eq. (9.13). An example of a multimode state generated from the vacuum is

$$|n_1, \ n_2, \ \ldots, \ n_l, \ \ldots\rangle = \left[\frac{\left(\widehat{a}_1^\dagger\right)^{n_1}}{\sqrt{n_1!}} \frac{\left(\widehat{a}_2^\dagger\right)^{n_2}}{\sqrt{n_2!}} \ldots \frac{\left(\widehat{a}_l^\dagger\right)^{n_l}}{\sqrt{n_l!}} \ldots\right]$$
$$\times |0_1, \ 0_2, \ \ldots, \ 0_l, \ \ldots\rangle \tag{9.14}$$

Equation (9.14) denotes a multimode quantized state with n_1 photons in mode 1 of energy $\hbar\omega_1$ and momentum $\hbar k_1 = \hbar\omega_1/c$, n_2 photons in mode 2 of energy $\hbar\omega_2$ and momentum $\hbar k_2 = \hbar\omega_2/c$, etc. Each mode is generated from the corresponding vacuum state $|0_1\rangle$, $|0_2\rangle$, etc. Equation (9.13) assumes that the state is factorizable. There are also entangled states, which cannot be factorized in this manner. An example of a multimode entangled state will be given in Sect. 9.4.

The multimode number operator is

$$\widehat{N} = \sum_l \widehat{a}_l^\dagger \widehat{a}_l \tag{9.15}$$

where

$$\widehat{N}|n_1, \ n_2, \ \ldots\rangle = (n_1 + n_2 + \ldots)|n_1, \ n_2, \ \ldots\rangle = \sum_l n_l|n_1, \ n_2, \ \ldots\rangle \tag{9.16}$$

$\sum_l n_l$ represents the total number of photons present in the state $|n_1, n_2, \ldots\rangle$.

9.3 Vacuum Energy

Let us apply the Hamiltonian, Eq. (9.3), to the vacuum state $|0_1, \ldots, 0_l, \ldots\rangle$:

$$\widehat{H}|0_1, \ \ldots, \ 0_l, \ \ldots\rangle = \frac{1}{2}\hbar\sum_l \omega_l|0_1, \ \ldots, \ 0_l, \ \ldots\rangle \qquad (9.17)$$

which gives the total energy

$$E = \frac{1}{2}\hbar\sum_l \omega_l \qquad (9.18)$$

where $l = 1, 2, \cdots$ corresponding to each of the modes. Since l spans all integers, the vacuum energy is infinite! As we saw in Chap. 4, the vacuum has real consequences in the presence of matter or boundary conditions (such as the Casimir effect). Richard Feynman, Julian Schwinger, and Sin-Itiro Tomonaga were awarded the 1965 Nobel Prize in Physics for developing a sophisticated method of "renormalization" used to deal with the infinity of the vacuum energy in calculations. However, we usually do not need to worry about the infinite vacuum energy, since we measure the photon energy relative to the vacuum level; that is, we can only measure energy differences.

9.4 Single Photon Wavepacket

A single photon state is an eigenstate of $\widehat{N}$ with eigenvalue of 1, which could be a multimode state. An example of a single photon multimode state is the superposition

$$|\psi\rangle = \frac{1}{\sqrt{2}}(|1_1\rangle + |1_2\rangle) \qquad (9.19)$$

which is a single photon in mode 1 with frequency ω_1 and in mode 2 with frequency ω_2. Note that

$$\begin{aligned}
\widehat{N}|\psi\rangle &= \sum_l \widehat{a}_l^\dagger \widehat{a}_l|\psi\rangle \\
&= \frac{1}{\sqrt{2}}\left(\widehat{a}_1^\dagger \widehat{a}_1|1_1\rangle + \widehat{a}_2^\dagger \widehat{a}_2|1_2\rangle\right) \\
&= \frac{1}{\sqrt{2}}(1|1_1\rangle + 1|1_2\rangle) \\
&= 1|\psi\rangle
\end{aligned} \qquad (9.20)$$

Thus, $|\psi\rangle$ is an eigenstate of $\widehat{N}$ with eigenvalue 1; that is, by definition, $|\psi\rangle$ is a single photon state.

Exercise 9.1 Show that the single photon state given by Eq. (9.19) is not an eigenstate of the Hamiltonian.

In general, a one photon wavepacket is given by

$$|\psi\rangle = \sum_l c_l |1_l\rangle \qquad (9.21)$$

which is an eigenstate of $\widehat{N}$ with eigenvalue 1. The coefficients c_l must satisfy the normalization condition:

$$\sum_l |c_l|^2 = 1 \qquad (9.22)$$

9.5 Spontaneous Emission

As an example of the application of multimode radiation, we evaluate the detection of a single photon wavepacket emitted by spontaneous emission. A single photon wavepacket can be produced by spontaneous emission of a photon due to transition of an electron from an excited state $|e\rangle$ to a ground state $|g\rangle$ of an atom, as illustrated in Fig. 9.1a. We suppose the electron was prepared in the excited state at time $t = 0$ and has a lifetime γ^{-1}. After some time $t_0 \gg \gamma^{-1}$, a transition of the electron occurs from the excited state to the ground state accompanied by the emission of a single photon wavepacket. Here, it is useful to use the Heisenberg picture of the time-dependence (Sect. 3.5). Like Eq. (9.21), the single photon wavepacket emitted at some time t_0, using the Heisenberg picture, is given by

$$|\psi(t_0)\rangle = \sum_l c_l |1_l\rangle \qquad (9.23)$$

Fig. 9.1 (a) Energy diagram and (b) Lorentzian distribution of frequencies in a multimode single photon wavepacket produced by spontaneous emission

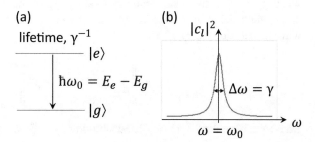

We suppose the emission from an atomic transition, like Fig. 9.1a, has a Lorentzian distribution of frequencies, centered at $\omega_0 = (E_e - E_g)/\hbar$ and with width γ, as shown in Fig. 9.1b (Exercise 9.2). The Lorentzian distribution of frequencies is due to the finite lifetime, γ^{-1}, of the excited state. According to the energy-time uncertainty relation, a finite lifetime γ^{-1} produces a spread γ in frequency (hence, a multimode state). The probability amplitudes c_l of the modes l with frequency ω_l are described by

$$c_l = \frac{K}{(\omega_l - \omega_0) + i\gamma/2} \tag{9.24}$$

where $\Delta\omega = \gamma$ is the width of the frequency distribution. Eq. (9.24) gives a Lorentzian function for the probabilities $|c_l|^2$ of the modes l:

$$|c_l|^2 = \frac{K^2}{(\omega_l - \omega_0)^2 + \gamma^2/4} \tag{9.25}$$

Here, K is a normalization constant used to satisfy the normalization condition in Eq. (9.22). K is derived in Appendix 3. Eq. (9.25) gives the natural linewidth of the spontaneously emitted radiation in the absence of other broadening mechanisms.

Exercise 9.2 Why is the distribution of frequencies from a lifetime-limited spontaneous emission described by a Lorentzian function?

Using the Heisenberg picture, the photodetector signal (number of photon counts), I, like Eq. (5.9), is proportional to

$$I \propto \left\| \hat{E}^+ (r)|\psi\rangle \right\|^2 \tag{9.26}$$

$$\propto \left\| \sum_l \hat{a}_l e^{i[k_l z - \omega_l(t-t_0)]} |\psi(t_0)\rangle \right\|^2 \tag{9.27}$$

where $t_0 + z/c$ is the delay time due to emission of the wavepacket at time t_0 and propagation delay over distance z from the source to the detector. Substituting the wavepacket, Eq. (9.23), gives

$$I \propto \left\| \sum_l c_l e^{i[k_l z - \omega_l(t-t_0)]} |0\rangle \right\|^2 \tag{9.28}$$

where we have used $\hat{a}|1\rangle = |0\rangle$. Substituting $k_l = \frac{\omega_l}{c}$ and defining $\tau = t - t_0 - z/c$ gives

$$I \propto \left\| \sum_l c_l e^{-i\omega_l \tau} |0\rangle \right\|^2 \tag{9.29}$$

Substituting Eq. (9.24) gives

$$I \propto \left\| \sum_l \frac{K e^{-i\omega_l \tau}}{(\omega_l - \omega_0) + i\gamma/2} |0\rangle \right\|^2 \tag{9.30}$$

For a continuous frequency distribution, Eq. (9.30) becomes

$$I \propto \left\| K e^{-i\omega_0 \tau} \left(\int \frac{e^{-i\Omega\tau}}{\Omega + i\gamma/2} d\Omega \right) |0\rangle \right\|^2 \tag{9.31}$$

where $\Omega = \omega_l - \omega_0$. The integral in Eq. (9.31) is the Fourier transform of a Lorentzian. The integration results in

$$I \propto \left\| K e^{-i\omega_0 \tau} \left(-2\pi i H(\tau) e^{-\gamma\tau/2} \right) |0\rangle \right\|^2 \tag{9.32}$$

where $H(\tau)$ is the Heaviside or step function, shown in Fig. 9.2. Finally, the photodetection signal is proportional to

Fig. 9.2 Heaviside or step function, $H(\tau)$

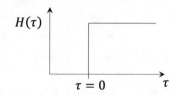

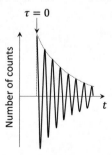

Fig. 9.3 Photodetector signal, I (number of photon counts) versus time for a single photon wavepacket. The dashed line is the envelope of the wavepacket, representing the number of counts measured at a photodetector. The electric field oscillations around frequency ω_0, shown within the envelope, are too rapid in the optical domain to be directly detected (there would be many more oscillations than depicted here)

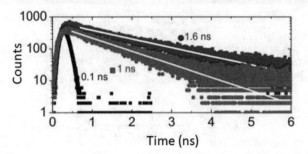

Fig. 9.4 Time-correlated single photon counting (TCSPC) from a single indium arsenide (InAs) quantum dot. Blue circles correspond to emission due to exciton recombination and red squares to biexciton recombination. Black squares are the instrument response. The single exponential fit, like Eq. (9.33), is shown by the solid white lines giving γ^{-1} in nanoseconds. (Reprinted by permission from Springer Nature, M. Birowosuto et al. [1])

$$I \propto H(\tau)e^{-\gamma\tau} \tag{9.33}$$

Thus, the expected photodetector signal is a sharp increase followed by an exponential decay as shown by the dashed line in Fig. 9.3a. The time-dependence described by Eq. (9.33) is observed experimentally as shown, for example, in Fig. 9.4 for indium arsenide quantum dots measured using "time-correlated single photon counting" (TCSPC). The photodetector signal is obtained as a histogram from many single photon detection events as a function of time.

Reference

1. M. Birowosuto et al., *Fast Purcell-enhanced single photon source in 1,550-nm telecom band from a resonant quantum dot-cavity coupling*, Sci. Rep. 2 (2012) 321.

Chapter 10
Coherent State

A coherent state, such as that from a laser, is not just a bunch of photons. How do we describe a state that is a coherent wave like that from a laser? We will see that the coherent state can be described as a superposition of Fock states. The properties of the coherent state are derived, including the Poisson distribution of photon number. The quadrature components of the coherent state are derived, showing that the coherent state is a minimum uncertainty state, leading to the shot noise limit. The phasor representation of the electric field is introduced, and the number-phase uncertainty relation is derived.

10.1   Coherent State

The definition of a coherent state is

$$\hat{a}|\alpha\rangle = \alpha|\alpha\rangle \tag{10.1}$$

where the eigenvalue, α, is a complex number ($\alpha \in \mathbb{C}$) given by

$$\alpha = |\alpha|e^{i\varphi} \tag{10.2}$$

We are using the standard notation where α labels both the eigenstate and the eigenvalue. The Hermitian conjugate of Eq. (10.1) gives

$$\langle \alpha|\hat{a}^\dagger = \langle \alpha|\alpha^* \tag{10.3}$$

According to Eq. (10.1), a coherent state is an eigenstate of the annihilation operator with eigenvalue α. This means that a coherent state is also an eigenstate of the field operator $\widehat{E}^+$ with eigenvalue proportional to α. Thus, unlike Fock states that

© The Author(s), under exclusive license to Springer Nature Switzerland AG 2022
R. LaPierre, *Getting Started in Quantum Optics*, Undergraduate Texts in Physics,
https://doi.org/10.1007/978-3-031-12432-7_10

have average electric field equal to zero ($\langle E \rangle = 0$), we will see that Eq. (10.1) means that the expectation value of the electric field will not vanish.

10.2   Coherent State as a Superposition of Fock States

It is useful to express $|\alpha\rangle$ in terms of the basis states $|n\rangle$ (superposition of single mode Fock states):

$$|\alpha\rangle = e^{-|\alpha|^2/2} \sum_{n=0}^{\infty} \frac{\alpha^n}{\sqrt{n!}} |n\rangle \tag{10.4}$$

We can check that Eq. (10.4) is correct by verifying that it satisfies the eigenvalue equation, Eq. (10.1):

$$\hat{a}|\alpha\rangle = e^{-|\alpha|^2/2} \sum_{n=0}^{\infty} \frac{\alpha^n}{\sqrt{n!}} \hat{a}|n\rangle \tag{10.5}$$

Using Eq. (2.127), $\hat{a}|n\rangle = \sqrt{n}|n-1\rangle$, gives

$$\hat{a}|\alpha\rangle = e^{-|\alpha|^2/2} \sum_{n=1}^{\infty} \frac{\alpha^n}{\sqrt{n!}} \sqrt{n}|n-1\rangle \tag{10.6}$$

Note that the index in Eq. (10.5) starts at $n = 0$, while the index in (10.6) starts at $n = 1$, since the lowest possible state is $|0\rangle$. Since $\frac{\sqrt{n}}{\sqrt{n!}} = \frac{1}{\sqrt{(n-1)!}}$, we get

$$\hat{a}|\alpha\rangle = e^{-|\alpha|^2/2} \sum_{n=1}^{\infty} \frac{\alpha^n}{\sqrt{(n-1)!}} |n-1\rangle \tag{10.7}$$

$$= \alpha \left(e^{-|\alpha|^2/2} \sum_{n=1}^{\infty} \frac{\alpha^{n-1}}{\sqrt{(n-1)!}} |n-1\rangle \right) \tag{10.8}$$

We can change the starting index in Eq. (10.8) back to 0, giving the term in brackets as $|\alpha\rangle = e^{-|\alpha|^2/2} \sum_{n=0}^{\infty} \frac{\alpha^n}{\sqrt{n!}} |n\rangle$, which is identical to Eq. (10.4). Thus, Eq. (10.8) becomes

$$\hat{a}|\alpha\rangle = \alpha|\alpha\rangle \tag{10.9}$$

which is Eq. (10.1).

Is Eq. (10.4) normalized? We need to verify that $\langle \alpha | \alpha \rangle = 1$:

$$\langle \alpha | \alpha \rangle = e^{-|\alpha|^2} \sum_{n,n'=0}^{\infty} \frac{(\alpha^*)^n (\alpha)^{n'}}{\sqrt{n!}\sqrt{n'!}} \langle n | n' \rangle \tag{10.10}$$

The states $|n\rangle$ and $|n'\rangle$ are orthonormal; that is, $\langle n | n' \rangle = \delta_{n,n'}$, meaning that each term in the summation is nonzero only if $n = n'$, which gives

$$\langle \alpha | \alpha \rangle = e^{-|\alpha|^2} \sum_{n=0}^{\infty} \frac{|\alpha|^{2n}}{n!} \tag{10.11}$$

The summation in Eq. (10.11) contains the well-known expansion for the exponential function:

$$e^{|\alpha|^2} = \sum_{n=0}^{\infty} \frac{|\alpha|^{2n}}{n!} \tag{10.12}$$

Therefore, from Eq. (10.11),

$$\langle \alpha | \alpha \rangle = e^{-|\alpha|^2} e^{|\alpha|^2} = 1 \tag{10.13}$$

Thus, we have proven that the state $|\alpha\rangle$, as written in Eq. (10.4), is normalized.

Exercise 10.1 Show that two coherent states satisfy $|\langle \alpha | \beta \rangle|^2 = e^{-|\alpha - \beta|^2}$. We say that two coherent states are "quasiorthogonal"; that is, the two states become increasingly orthogonal with the separation of α and β in the complex plane.

10.3 Photon Number

The average photon number of a coherent state is

$$\langle n \rangle = \langle \alpha | \widehat{N} | \alpha \rangle \tag{10.14}$$

$$= \langle \alpha | \widehat{a}^\dagger \widehat{a} | \alpha \rangle \tag{10.15}$$

$$= \langle \alpha | \alpha^* \alpha | \alpha \rangle \tag{10.16}$$

$$= |\alpha|^2 \tag{10.17}$$

Note that we can interpret Eq. (10.17) as giving a connection between the particle (photon) view and the wave view. $\langle n \rangle$ is the average photon number (particle view), while $|\alpha|^2$ is proportional to the square of a field amplitude (the wave intensity).

Next, let us calculate the uncertainty in n. To do so, we first calculate $\langle n^2 \rangle$:

$$\langle n^2 \rangle = \langle \alpha | \hat{a}^\dagger \hat{a}\, \hat{a}^\dagger \hat{a} | \alpha \rangle \tag{10.18}$$

Using Eqs. (10.1) and (10.3), we get

$$\langle n^2 \rangle = |\alpha|^2 \langle \alpha | \hat{a}\, \hat{a}^\dagger | \alpha \rangle \tag{10.19}$$

Putting this in the normal order gives

$$\langle n^2 \rangle = |\alpha|^2 \langle \alpha | (1 + \hat{a}^\dagger \hat{a}) | \alpha \rangle \tag{10.20}$$

$$= |\alpha|^2 \left(1 + |\alpha|^2 \right) \tag{10.21}$$

$$= |\alpha|^2 + |\alpha|^4 \tag{10.22}$$

$$= \langle n \rangle + \langle n \rangle^2 \tag{10.23}$$

The uncertainty in n is

$$\Delta n = \sqrt{\langle n^2 \rangle - \langle n \rangle^2} \tag{10.24}$$

Using Eq. (10.23), we get

$$\Delta n = \sqrt{\langle n \rangle + \langle n \rangle^2 - \langle n \rangle^2} \tag{10.25}$$

$$= \sqrt{\langle n \rangle} \tag{10.26}$$

Equation (10.26) is known as the shot noise limit. Note that, unlike the Fock state, a coherent state is not an eigenstate of the number operator, since there is a dispersion (uncertainty) in the photon number. Also note that the coherent state is not an eigenstate of the Hamiltonian. However, it is easy to calculate the average energy, $\langle \alpha | \hat{H} | \alpha \rangle = \langle \alpha | \hbar\omega\, (\hat{a}^\dagger \hat{a} + \tfrac{1}{2}) | \alpha \rangle = \hbar\omega \big(\langle n \rangle + \tfrac{1}{2} \big)$.

10.4   Poisson Distribution

In Eq. (10.4), we expressed the coherent state as a superposition of Fock states:

$$|\alpha\rangle = \sum_{n=0}^{\infty} c_n |n\rangle = e^{-|\alpha|^2/2} \sum_{n=0}^{\infty} \frac{\alpha^n}{\sqrt{n!}} |n\rangle \tag{10.27}$$

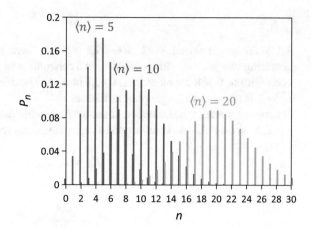

Fig. 10.1 Poisson distribution P_n for average photon number $\langle n \rangle = 5$, 10 and 20

We are interested in the probability distribution of the photon number, $P_n = |c_n|^2$:

$$P_n = |c_n|^2 = e^{-|\alpha|^2} \frac{|\alpha|^{2n}}{n!} \tag{10.28}$$

or, using Eq. (10.17):

$$P_n = e^{-\langle n \rangle} \frac{\langle n \rangle^n}{n!} \tag{10.29}$$

Equation (10.29) is the Poisson probability distribution! P_n is the probability of detecting n photons in a fixed time interval if $\langle n \rangle$ is the average number of photons in that time interval. The Poisson distribution, P_n, is shown in Fig. 10.1 for various average values $\langle n \rangle$. You can see that the spread (uncertainty) increases as n increases. It can be shown that, as $\langle n \rangle$ increases, the Poisson distribution approaches a Gaussian distribution with standard deviation of $\sqrt{\langle n \rangle}$ [1]:

$$\underbrace{P_n = e^{-\langle n \rangle} \frac{\langle n \rangle^n}{n!}}_{\text{Poisson distribution}} \xrightarrow{\text{large } n} \underbrace{P_n = \frac{1}{\sqrt{2\pi \langle n \rangle}} e^{-\frac{(n - \langle n \rangle)^2}{2\langle n \rangle}}}_{\text{Gaussian distribution}} \tag{10.30}$$

Suppose we have a source with $\langle n \rangle = 1$. The probability of observing 0, 1, 2 or 3 photons from Eq. (10.29) is (note that the factorial of 0 is equal to 1)

$$P_0 = 0.368 \tag{10.31}$$

$$P_1 = 0.368 \tag{10.32}$$

$$P_2 = 0.184 \tag{10.33}$$

$$P_3 = 0.061 \tag{10.34}$$

Unlike the single photon Fock state with $\langle n \rangle = 1$, there is a probability of 0.184 of measuring two photons simultaneously in a coherent state with $\langle n \rangle = 1$, and an equal probability of 0.368 for measuring 0 or 1 photon. Therefore, a coherent source with $\langle n \rangle = 1$ is not a good single photon source!

Suppose we attenuate a coherent source (e.g., attenuate a laser beam), such that $\langle n \rangle \ll 1$. Would this produce a good single photon source? Equation (10.29) gives

$$P_0 \sim 1 \tag{10.35}$$

$$P_1 \sim \langle n \rangle \tag{10.36}$$

$$P_2 \sim \frac{\langle n \rangle^2}{2} \tag{10.37}$$

P_0 is less than but close to 1. This means that a measurement would result in zero photons most of the time. A single photon would be detected with probability $\langle n \rangle$, two photons with probability $\frac{\langle n \rangle^2}{2}$, etc. Note that the sum of probabilities will yield 1 if we continue with the higher-order terms. For $\langle n \rangle \ll 1$, like in a strongly attenuated laser source, there is still a finite (albeit low) probability for double photon detection, and most of the time zero photons will be measured in a given time interval. You cannot produce an efficient single photon source by attenuating a coherent source!

10.5   Electric Field of Coherent State

Recall that the electric field operator is

$$\widehat{E}(r) = i\varepsilon\varepsilon^1 \left(\widehat{a} e^{ik \cdot r} - \widehat{a}^\dagger e^{-ik \cdot r} \right) \tag{10.38}$$

The average electric field for a coherent state is

$$\langle E(r) \rangle = i\varepsilon\varepsilon^1 \langle \alpha | (\widehat{a} e^{ik \cdot r} - \widehat{a}^\dagger e^{-ik \cdot r}) | \alpha \rangle \tag{10.39}$$

$$= i\varepsilon\varepsilon^1 \left(\alpha e^{ik \cdot r} - \alpha^* e^{-ik \cdot r} \right) \tag{10.40}$$

$$= i\varepsilon\varepsilon^1 \alpha e^{ik \cdot r} + c.c. \tag{10.41}$$

If we include the time-dependence (see Sect. 10.7), we get

$$\langle E(r, t) \rangle = i\varepsilon\varepsilon^1 \alpha e^{i(k \cdot r - \omega t)} + c.c. \tag{10.42}$$

Using Eq. (10.2), we get

$$\langle \boldsymbol{E}(\boldsymbol{r},\ t)\rangle = -2\varepsilon\varepsilon^1|\alpha|\sin\left(\boldsymbol{k}\cdot\boldsymbol{r} - \omega t + \varphi\right) \tag{10.43}$$

Equation (10.43) looks like a classical monochromatic traveling plane wave and is different than the number (Fock) state where $\langle \boldsymbol{E}\rangle = 0$. The coherent state is the quantum description of light that most closely resembles classical coherent light, such as that from a laser. However, unlike the classical description of light, the electric field of the coherent state has an uncertainty that we calculate below. For this reason, coherent states are often called "quasiclassical" states.

As usual, let us find the uncertainty in electric field, ΔE, by first calculating $\langle \boldsymbol{E}^2\rangle$. Using Eq. (10.38):

$$\langle \boldsymbol{E}^2\rangle = \langle \alpha|\widehat{\boldsymbol{E}}^2|\alpha\rangle \tag{10.44}$$

$$= -\left(\varepsilon^1\right)^2 \langle \alpha|(\widehat{a}e^{i\boldsymbol{k}\cdot\boldsymbol{r}} - \widehat{a}^\dagger e^{-i\boldsymbol{k}\cdot\boldsymbol{r}})(\widehat{a}e^{i\boldsymbol{k}\cdot\boldsymbol{r}} - \widehat{a}^\dagger e^{-i\boldsymbol{k}\cdot\boldsymbol{r}})|\alpha\rangle \tag{10.45}$$

$$= -\left(\varepsilon^1\right)^2\langle \alpha|(\widehat{a}^2 e^{2i\boldsymbol{k}\cdot\boldsymbol{r}} - \widehat{a}\widehat{a}^\dagger - \widehat{a}^\dagger\widehat{a} + \left(\widehat{a}^\dagger\right)^2 e^{-2i\boldsymbol{k}\cdot\boldsymbol{r}})|\alpha\rangle \tag{10.46}$$

$$= -\left(\varepsilon^1\right)^2\langle \alpha|(\widehat{a}^2 e^{2i\boldsymbol{k}\cdot\boldsymbol{r}} - (1 + \widehat{a}^\dagger\widehat{a}) - \widehat{a}^\dagger\widehat{a} + \left(\widehat{a}^\dagger\right)^2 e^{-2i\boldsymbol{k}\cdot\boldsymbol{r}})|\alpha\rangle \tag{10.47}$$

$$= -\left(\varepsilon^1\right)^2\left[(\alpha)^2 e^{2i\boldsymbol{k}\cdot\boldsymbol{r}} - 1 - 2\alpha^*\alpha + (\alpha^*)^2 e^{-2i\boldsymbol{k}\cdot\boldsymbol{r}}\right] \tag{10.48}$$

From Eqs. (10.41) and (10.48), we get

$$\langle \boldsymbol{E}^2\rangle = \langle \boldsymbol{E}\rangle^2 + \left(\varepsilon^1\right)^2 \tag{10.49}$$

Using Eq. (10.49), the uncertainty ΔE is

$$\Delta E = \sqrt{\langle \boldsymbol{E}^2\rangle - \langle \boldsymbol{E}\rangle^2} \tag{10.50}$$

$$= \sqrt{\langle \boldsymbol{E}\rangle^2 + \left(\varepsilon^1\right)^2 - \langle \boldsymbol{E}\rangle^2} \tag{10.51}$$

$$= \varepsilon^1 \tag{10.52}$$

Thus, according to our discussion in Chap. 4, a coherent state is a minimum uncertainty state with field fluctuations ε^1.

10.6 Phasor Representation

Choosing $r = 0$ for simplicity, we can write Eq. (10.43) as

$$\langle E(t) \rangle = -2\varepsilon\varepsilon^1 |\alpha| \sin(-\omega t + \varphi) \tag{10.53}$$

Using Eq. (10.53), $-\langle E(t) \rangle$ can be represented as a rotating phasor (rotating point on a circle in the complex plane), as shown in Fig. 10.2a, with amplitude $2\varepsilon^1 |\alpha|$, initial phase φ at $t = 0$, and rotating clockwise at rate ω. Note the negative sign in the amplitude of Eq. (10.53). Thus, $-\langle E(t) \rangle$ is given by the projection onto the imaginary axis, as shown in Fig. 10.2b. According to Eq. (10.30), the gray circle in Fig. 10.2a represents a Gaussian distribution (as depicted in Fig. 10.1) in the limit of large $\langle n \rangle = |\alpha|^2$; that is, large phasor amplitude. The standard deviation $\Delta E = \varepsilon^1$ of the Gaussian distribution gives uncertainty in the field of $\pm \varepsilon^1$ around the average, and a width of $2\varepsilon^1$ represented by the diameter of the gray circle in Fig. 10.2a. The signal-to-noise ratio (SNR) may be given by

$$\text{SNR} = \frac{2|\alpha|\varepsilon^1}{2\varepsilon^1} = |\alpha| = \sqrt{\langle n \rangle} \tag{10.54}$$

Classical optics (including laser light) can be explained as a macroscopic limit of quantum optics. For example, suppose a laser contains $\langle n \rangle \sim 10^{10}$ photons in its cavity. The uncertainty in photon number from Eq. (10.26) is $\sqrt{\langle n \rangle} \sim 10^5$, which is much less than $\langle n \rangle$. As α (or $\langle n \rangle$) increases, the field becomes more classical. Note

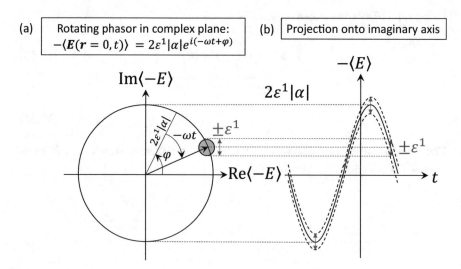

Fig. 10.2 (a) Phasor representation of the field $-\langle E(t) \rangle$ for the coherent state. (b) Projection of the rotating phasor onto the imaginary axis gives $-\langle E(t) \rangle$ with uncertainty ε^1 represented by the dashed lines and red arrows

that the electric field uncertainty ε^1 remains constant, but becomes negligible in comparison to the amplitude for large electric fields; that is, the gray circle in Fig. 10.2a becomes a relatively small rotating point and approximates the phasor for a classical wave.

10.7 Time-Dependence of Coherent State

Repeating Eq. (10.27), the coherent state is a superposition of Fock states:

$$|\alpha\rangle = \sum_{n=0}^{\infty} c_n |n\rangle = e^{-|\alpha|^2/2} \sum_{n=0}^{\infty} \frac{\alpha^n}{\sqrt{n!}} |n\rangle \tag{10.55}$$

The Fock states are an eigenstate of the Hamiltonian with eigenvalue $E_n = n\hbar\omega$ (ignoring the vacuum energy). We know the time-dependence of the Fock states in the Schrodinger picture is $|n(t)\rangle = e^{-iE_n t/\hbar} |n\rangle$. Thus, the time-dependence of the coherent state is given by

$$|\alpha(t)\rangle = \sum_{n=0}^{\infty} c_n e^{-iE_n t/\hbar} |n\rangle \tag{10.56}$$

or, using Eq. (10.55), we get

$$|\alpha(t)\rangle = e^{-|\alpha|^2/2} \sum_{n=0}^{\infty} \frac{\alpha^n}{\sqrt{n!}} e^{-in\omega t} |n\rangle \tag{10.57}$$

The time-dependence of the state from Eq. (10.57) can be written succinctly as

$$|\alpha(t)\rangle = |\alpha(0)e^{-i\omega t}\rangle \tag{10.58}$$

or, in terms of the eigenvalue:

$$\alpha(t) = \alpha(0) \, e^{-i\omega t} \tag{10.59}$$

We see that a coherent state remains a coherent state for all time. However, a coherent state is not an eigenstate of the Hamiltonian. Therefore, a coherent state evolves in time as shown in Fig. 10.2.

Let us find the time-dependent expectation values for the Q and P quadratures. Recalling Eqs. (2.44) and (2.45), in the Heisenberg picture, we get

$$Q = \langle Q \rangle = \langle \alpha | \hat{Q} | \alpha \rangle = \frac{1}{\sqrt{2}} \langle \alpha | (\hat{a} + \hat{a}^\dagger) | \alpha \rangle = \frac{1}{\sqrt{2}} \left(\alpha e^{-i\omega t} + \alpha^* e^{i\omega t} \right) \tag{10.60}$$

Using Eq. (10.2), we get

$$Q = \frac{1}{\sqrt{2}} \left(|\alpha| e^{i(-\omega t + \varphi)} + |\alpha| e^{-i(-\omega t + \varphi)} \right) \tag{10.61}$$

$$= \sqrt{2} |\alpha| \cos(-\omega t + \varphi) \tag{10.62}$$

Similarly,

$$P = \langle P \rangle = \langle \alpha | \widehat{P} | \alpha \rangle \tag{10.63}$$

$$= \frac{-i}{\sqrt{2}} \langle \alpha | (\widehat{a} - \widehat{a}^{\dagger}) | \alpha \rangle$$

$$= \frac{-i}{\sqrt{2}} \left(\alpha e^{-i\omega t} - \alpha^* e^{i\omega t} \right)$$

$$= \frac{-i}{\sqrt{2}} \left(|\alpha| e^{i(-\omega t + \varphi)} - |\alpha| e^{-i(-\omega t + \varphi)} \right)$$

$$= \sqrt{2} |\alpha| \sin(-\omega t + \varphi) \tag{10.64}$$

Recalling the classical harmonic oscillator from Chap. 2, Q represented the dimensionless position and P represented the dimensionless momentum. We see that the expectation values, Q and P for the electric field, oscillate with time and are 90° out of phase with each other just like the position and momentum of a classical harmonic oscillator. It is also in this sense that coherent states are the "most classical of states", since they are analogous to the dynamics of a classical harmonic oscillator. The coherent state produces an oscillating Gaussian wavepacket (Fig. 10.3, red), analogous to a classical particle oscillating in a parabolic potential. If the wavepacket is to reproduce a classical particle, we better make sure that its width is not changing with position or time as it oscillates. In the next section, we show that the Gaussian wavepacket of the coherent state is a minimum uncertainty state whose width (ΔQ, ΔP) remains the same for all time.

Fig. 10.3 A particle oscillating in a parabolic potential, $U(x)$. The quantum analogue is a Gaussian wavepacket (red) representing the particle position (Q quadrature), while the momentum is represented by the P quadrature

10.8   Quadratures

In the Schrodinger picture (set $t = 0$ in Eqs. (10.62) and (10.64)), we have

$$Q = \sqrt{2}|\alpha| \cos \varphi = \sqrt{2} \, \text{Re} \, (\alpha) = \frac{1}{\sqrt{2}}(\alpha + \alpha^*) \tag{10.65}$$

and

$$P = \sqrt{2} \, |\alpha| \sin \varphi = \sqrt{2} \, \text{Im}(\alpha) = \frac{-i}{\sqrt{2}}(\alpha - \alpha^*) \tag{10.66}$$

Next, let us find the uncertainties, starting with $\langle Q^2 \rangle$

$$\langle Q^2 \rangle = \langle \alpha | \hat{Q}^2 | \alpha \rangle = \frac{1}{2} \langle \alpha | (\hat{a} + \hat{a}^\dagger)(\hat{a} + \hat{a}^\dagger) | \alpha \rangle \tag{10.67}$$

$$= \frac{1}{2} \langle \alpha | (\widehat{aa} + \widehat{aa}^\dagger + \hat{a}^\dagger \hat{a} + \hat{a}^\dagger \hat{a}^\dagger) | \alpha \rangle \tag{10.68}$$

Using the commutation relation, $\widehat{aa}^\dagger = 1 + \hat{a}^\dagger \hat{a}$, for normal ordering gives

$$\langle Q^2 \rangle = \frac{1}{2} \langle \alpha | (\widehat{aa} + 1 + 2\hat{a}^\dagger \hat{a} + \hat{a}^\dagger \hat{a}^\dagger) | \alpha \rangle \tag{10.69}$$

$$= \frac{1}{2} \left((\alpha)^2 + 1 + 2|\alpha|^2 + (\alpha^*)^2 \right) \tag{10.70}$$

$$= \frac{1}{2} \left[(\alpha + \alpha^*)^2 + 1 \right] \tag{10.71}$$

Thus, the uncertainty is

$$\Delta Q = \sqrt{\langle Q^2 \rangle - \langle Q \rangle^2} \tag{10.72}$$

$$= \sqrt{\frac{1}{2} \left[(\alpha + \alpha^*)^2 + 1 \right] - \left[\frac{1}{\sqrt{2}}(\alpha + \alpha^*) \right]^2} \tag{10.73}$$

$$= \frac{1}{\sqrt{2}} \tag{10.74}$$

Similarly,

$$\Delta P = \frac{1}{\sqrt{2}} \tag{10.75}$$

Thus,

$$\Delta Q \Delta P = \frac{1}{2} \tag{10.76}$$

Hence, there is an uncertainty relation between the electric field at a certain point in time and the electric field at a quarter cycle later. The coherent state is a minimum uncertainty state satisfying the standard quantum limit, like the ground state of the quantum harmonic oscillator. A coherent state distributes its quantum mechanical uncertainties equally between the Q and P quadratures.

Exercise 10.2 Derive Eq. (10.75).

The quadrature representation, shown in Fig. 10.4, represents α in the complex plane, according to Eq. (10.2), with magnitude (radius of the circle) given by $|\alpha| = \sqrt{\langle n \rangle}$ and with initial phase φ. The projections give the real and imaginary parts of α according to Eqs. (10.65) and (10.66). The gray circle represents what would be obtained from many different measurements of P and Q with spread given by Eqs. (10.74) and (10.75).

We remind the reader from Eq. (3.49) that the electric field can be written in terms of the quadrature operators as

Fig. 10.4 Quadrature representation of the coherent state

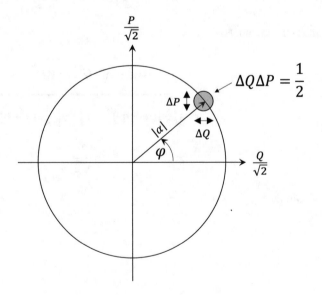

$$\widehat{E}(r, t) = -\varepsilon\varepsilon^{l}\sqrt{2}\left[\widehat{P}\cos{(k \cdot r - \omega t)} + \widehat{Q}\sin{(k \cdot r - \omega t)}\right] \quad (10.77)$$

giving

$$\langle E(r, \ t)\rangle = \langle\alpha|\widehat{E}(r, t)|\alpha\rangle \quad (10.78)$$

$$= -2\varepsilon\varepsilon^{l}|\alpha|\left[\sin\varphi\cos{(k \cdot r - \omega t)} + \cos\varphi\sin{(k \cdot r - \omega t)}\right] \quad (10.79)$$

Is there a way to fix φ, so that we can detect the quadrature components? In fact, there is! The method is known as homodyne detection, which is treated in Chap. 13.

10.9 Displacement Operator

Note that the vacuum state gave $\widehat{a}|0\rangle = 0$. Thus, comparing with Eq. (10.1), the vacuum state can be considered as a coherent state with $\alpha = 0$. It is useful to think of the coherent state as a "displaced" vacuum, as depicted in Fig. 10.5. The displacement can be performed by a displacement operator given by

$$\widehat{D}(\alpha) = e^{\left(\alpha\widehat{a}^{\dagger} - \alpha^{*}\widehat{a}\right)} \quad (10.80)$$

Fig. 10.5 Coherent state represented as a displaced vacuum

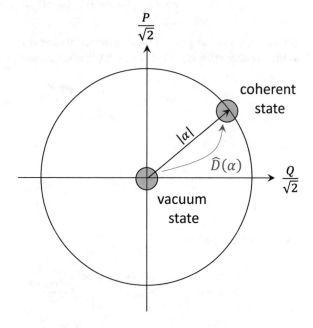

We can rewrite this operator using the following theorem for the exponential of operators, often called Glauber's formula:

$$e^{\widehat{A}+\widehat{B}} = e^{\widehat{A}}e^{\widehat{B}}e^{-\frac{1}{2}[\widehat{A},\,\widehat{B}]} \tag{10.81}$$

where $[\widehat{A}, \widehat{B}] \neq 0$ and providing that

$$\left[\widehat{A}, [\widehat{A}, \widehat{B}]\right] = \left[\widehat{B}, [\widehat{A}, \widehat{B}]\right] = 0 \tag{10.82}$$

The latter conditions are met with $\widehat{A} = \alpha\widehat{a}^{\dagger}$ and $\widehat{B} = -\alpha^*\widehat{a}$ from Eq. (10.80). In this case,

$$\left[\widehat{A}, \widehat{B}\right] = |\alpha|^2 \tag{10.83}$$

The (tedious!) proof of Eq. (10.81) can be done by the Taylor expansion of the exponentials.

Exercise 10.3 Prove Eqs. (10.82) and (10.83) for $\widehat{A} = \alpha\widehat{a}^{\dagger}$ and $\widehat{B} = -\alpha^*\widehat{a}$.

Using the above results, we get

$$\widehat{D}(\alpha) = e^{\left(\alpha\widehat{a}^{\dagger} - \alpha^*\widehat{a}\right)} = e^{-\frac{1}{2}|\alpha|^2} e^{\alpha\widehat{a}^{\dagger}} e^{-\alpha^*\widehat{a}} \tag{10.84}$$

which is just another form of the displacement operator. Let us evaluate $\widehat{D}(\alpha)$ applied to the vacuum state $|0\rangle$. First, using the Taylor expansion of the exponential gives

$$e^{-\alpha^*\widehat{a}}|0\rangle = \sum_{n=0}^{\infty} \frac{(-\alpha^*\widehat{a})^n}{n!}|0\rangle = |0\rangle \tag{10.85}$$

since $\widehat{a}^n|0\rangle = 0$, except for $n = 0$. Next,

$$e^{\alpha\widehat{a}^{\dagger}}|0\rangle = \sum_{n=0}^{\infty} \frac{\left(\alpha\widehat{a}^{\dagger}\right)^n}{n!}|0\rangle = \sum_{n=0}^{\infty} \frac{\alpha^n}{n!}\left(\widehat{a}^{\dagger}\right)^n|0\rangle \tag{10.86}$$

Using Eq. (2.97) gives

$$e^{\alpha\widehat{a}^{\dagger}}|0\rangle = \sum_{n=0}^{\infty} \frac{\alpha^n}{\sqrt{n!}}|n\rangle \tag{10.87}$$

Thus,

$$\widehat{D}(\alpha)|0\rangle = e^{-\frac{1}{2}|\alpha|^2} e^{\widehat{a\alpha^\dagger}} e^{-\widehat{\alpha^* a}}|0\rangle = e^{-\frac{1}{2}|\alpha|^2} \sum_{n=0}^{\infty} \frac{\alpha^n}{\sqrt{n!}}|n\rangle \tag{10.88}$$

which is just the number representation of the coherent state. Thus, the displacement operator takes the Gaussian wavepacket of the $|0\rangle$ state (see Eq. (2.57) and Fig. 2.1) and translates it, creating a coherent state with the same uncertainty. Similarly, within a global phase factor, $\widehat{D}(\beta)|\alpha\rangle = |\alpha + \beta\rangle$, which results in displacement of the coherent state itself.

Exercise 10.4 Show that $\widehat{D}(\beta)|\alpha\rangle = |\alpha + \beta\rangle$ within a phase factor.

10.10   Number-Phase Uncertainty Relation

As shown in Fig. 10.6, the quadrature spread also corresponds to a spread in amplitude, $\Delta\alpha$, and phase, $\Delta\varphi$. Using the small angle approximation, the spread in phase from Fig. 10.6 is

Fig. 10.6 Number–phase uncertainty relation in the complex plane (phasor representation of α)

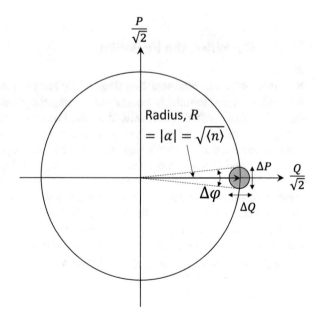

$$\Delta\varphi = \frac{\Delta P}{R} = \frac{\Delta P}{\sqrt{\langle n \rangle}} \qquad (10.87)$$

where we have used:

$$R^2 = |\alpha|^2 = \langle n \rangle \qquad (10.88)$$

Differentiating Eq. (10.88), we obtain the spread in photon number:

$$\Delta n = 2R\Delta R = 2\sqrt{\langle n \rangle}\Delta Q \qquad (10.89)$$

where we have used $R = \sqrt{\langle n \rangle}$ and $\Delta R = \Delta Q$ (see Fig. 10.6). Thus,

$$\Delta n\Delta\varphi = \left(\frac{\Delta P}{\sqrt{\langle n \rangle}}\right)\left(2\sqrt{\langle n \rangle}\Delta Q\right) = 2\Delta Q\Delta P \qquad (10.90)$$

Using Eq. (10.76), we obtain the number–phase uncertainty relation:

$$\Delta n\Delta\varphi \geq 1 \qquad (10.91)$$

As the photon number becomes larger, the phase becomes less uncertain, that is, we approach a classical state.

10.11   Revisiting the Fock State

Now that the phasor representation (Fig. 10.2a) has been introduced, we can revisit the Fock state and intuitively understand it using the phasor representation. Recall that the photon number is well defined for the Fock state with zero uncertainty. The average electric field of the Fock state is zero, but the uncertainty in electric field is nonzero. How can we reconcile these results in a phasor representation? A Fock state can be pictured as a superposition of many phasors, as depicted in Fig. 10.7a. The photon number, represented by the phasor amplitude, is well defined, but the phase angle is completely undefined, in accordance with the number–phase uncertainty relation. The resulting electric fields in Fig. 10.7b are given by the projection of the many phasors on the imaginary axis in Fig. 10.7a. The superposition of all these fields results in zero average electric field, $\langle E \rangle = 0$. However, the field fluctuations will be nonzero, such that $\langle E^2 \rangle \neq 0$ and thus $\Delta E \neq 0$.

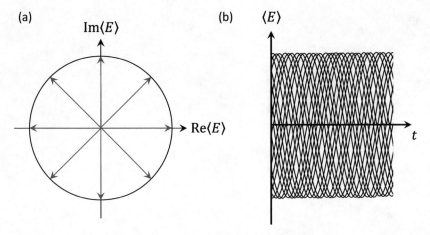

Fig. 10.7 (**a**) Fock state represented by a superposition of phasors (only a few are shown in red). (**b**) Resulting electric fields by projection of the phasors on the imaginary axis in (**a**), resulting in zero average electric field, $\langle E \rangle = 0$, but nonzero field fluctuations, $\langle E^2 \rangle \neq 0$ and $\Delta E \neq 0$

Reference

1. John R. Taylor, *An introduction to error analysis: The study of uncertainties in physical measurements* (2nd ed., Univ. Science Books, 1997).

Chapter 11
Coherent State on a Beam Splitter

In this short chapter, the photodetection probability of the coherent state on a beam splitter and the probability of coincidence measurements are derived. Unlike the single photon state, double photon counts (coincidences) occur for the coherent state, as we would expect for a classical coherent source like a laser.

11.1 Photodetection Probability

Let us consider a coherent state on a beam splitter, as shown in Fig. 11.1. The input state is $|\alpha\rangle_1|0\rangle_2$ with a coherent state on port 1 and "nothing" (vacuum) on port 2. The probability of detection at D_3 is

$$P_3 = \frac{\langle n_{\text{out}}\rangle}{\langle n_{\text{in}}\rangle} = \frac{\langle n_3\rangle}{|\alpha|^2} \tag{11.1}$$

where we have used $\langle n_{\text{in}}\rangle = |\alpha|^2$ for the coherent state input. Let us determine $\langle n_3\rangle$:

$$\langle n_3\rangle = \langle\psi_{\text{out}}|\hat{a}_3^\dagger\hat{a}_3|\psi_{\text{out}}\rangle \tag{11.2}$$

Writing Eq. (11.2) in terms of the input space, we get

$$\langle n_3\rangle = {}_2\langle 0|{}_1\langle\alpha|(r^*\hat{a}_1^\dagger + t^*\hat{a}_2^\dagger)(r\hat{a}_1 + t\hat{a}_2)|\alpha\rangle_1|0\rangle_2 \tag{11.3}$$

$$= {}_2\langle 0|{}_1\langle\alpha|(r^*r\hat{a}_1^\dagger\hat{a}_1 + r^*t\hat{a}_1^\dagger\hat{a}_2 + t^*r\hat{a}_2^\dagger\hat{a}_1 + t^*t\hat{a}_2^\dagger\hat{a}_2)|\alpha\rangle_1|0\rangle_2 \tag{11.4}$$

© The Author(s), under exclusive license to Springer Nature Switzerland AG 2022   111
R. LaPierre, *Getting Started in Quantum Optics*, Undergraduate Texts in Physics,
https://doi.org/10.1007/978-3-031-12432-7_11

Fig. 11.1 Coherent state on
a beam splitter

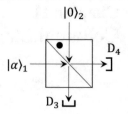

Remember that the operators act on the state with the same subscript. Thus,

$$_2\langle 0|_1\langle\alpha|\hat{a}_1^\dagger\hat{a}_2|\alpha\rangle_1|0\rangle_2 = 0 \tag{11.5}$$

$$_2\langle 0|_1\langle\alpha|\hat{a}_2^\dagger\hat{a}_1|\alpha\rangle_1|0\rangle_2 = 0 \tag{11.6}$$

$$_2\langle 0|_1\langle\alpha|\hat{a}_2^\dagger\hat{a}_2|\alpha\rangle_1|0\rangle_2 = 0 \tag{11.7}$$

Using Eqs. (11.5), (11.6) and (11.7), Eq. (11.4) becomes

$$\langle n_3\rangle = {}_2\langle 0|_1\langle\alpha|r^*r\,\hat{a}_1^\dagger\hat{a}_1|\alpha\rangle_1|0\rangle_2 \tag{11.8}$$

$$= r^*r\,\alpha\alpha^* = R|\alpha|^2 \tag{11.9}$$

Thus, from Eq. (11.1),

$$P_3 = R \tag{11.10}$$

which is the same as the classical result.

Exercise 11.1 Derive Eqs. (11.5), (11.6) and (11.7).

Similarly, the probability of detection at D_4 is

$$P_4 = \frac{\langle n_{\text{out}}\rangle}{\langle n_{\text{in}}\rangle} = \frac{\langle n_4\rangle}{|\alpha|^2} \tag{11.11}$$

where $\langle n_4\rangle$ is

$$\langle n_4\rangle = \langle\psi_{\text{out}}|\hat{a}_4^\dagger\hat{a}_4|\psi_{\text{out}}\rangle$$
$$= {}_2\langle 0|\,_1\langle\alpha|(t^*\hat{a}_1^\dagger - r^*\hat{a}_2^\dagger)(t\hat{a}_1 - r\hat{a}_2)|\alpha\rangle_1|0\rangle_2 \tag{11.12}$$

$$= {}_2\langle 0|_1 \langle \alpha| \left(t^* t \hat{a}_1^\dagger \hat{a}_1 - t^* r \hat{a}_1^\dagger \hat{a}_2 - r^* t \hat{a}_2^\dagger \hat{a}_1 + r^* r \hat{a}_2^\dagger \hat{a}_2 \right) |\alpha\rangle_1 |0\rangle_2 \qquad (11.13)$$

Keeping only the non-zero terms gives

$$\langle n_4 \rangle = {}_2\langle 0|_1 \langle \alpha| t^* t \, \hat{a}_1^\dagger \hat{a}_1 |\alpha\rangle_1 |0\rangle_2 \qquad (11.14)$$

$$= t^* t \, \alpha \alpha^* = T |\alpha|^2 \qquad (11.15)$$

Thus,

$$P_4 = T \qquad (11.16)$$

which is the same as the classical result.

11.2   Coincidence Measurements

The probability of simultaneous detection at D_3 and D_4 (correlation or coincidence measurement) is

$$\langle n_{34} \rangle = \langle \psi_{\text{out}} | \hat{a}_3^\dagger \hat{a}_4^\dagger \hat{a}_4 \hat{a}_3 | \psi_{\text{out}} \rangle \qquad (11.17)$$

Expressed in terms of the input space, we get

$$\langle n_{34} \rangle = {}_2\langle 0|_1 \langle \alpha| (r^* \hat{a}_1^\dagger + t^* \hat{a}_2^\dagger)(t^* \hat{a}_1^\dagger - r^* \hat{a}_2^\dagger)(t \hat{a}_1 - r \hat{a}_2)$$
$$\times (r \hat{a}_1 + t \hat{a}_2) |\alpha\rangle_1 |0\rangle_2 \qquad (11.18)$$

Note that, if we expand the terms, Eq. (11.18) is in the normal order. Retaining only the non-zero terms, we get

$$\langle n_{34} \rangle = {}_1\langle \alpha| (r^* \hat{a}_1^\dagger)(t^* \hat{a}_1^\dagger)(t \hat{a}_1)(r \hat{a}_1) |\alpha\rangle_1 \qquad (11.19)$$

$$= r^* t^* t r |\alpha|^4 \qquad (11.20)$$

$$= RT |\alpha|^4 \qquad (11.21)$$

Normalizing with $\langle n \rangle^2 = |\alpha|^4$, we get

$$P_{34} = RT \qquad (11.22)$$

which is the same as the classical result. This is what we would expect from a coherent source such as a laser.

The second-order correlation function, $g^{(2)}(0)$, is easily calculated. We have $P_3 = R$, $P_4 = T$, and $P_{34} = RT$. Thus,

$$g^{(2)}(0) = \frac{P_{34}}{P_3 P_4} = 1 \qquad (11.23)$$

Coherent sources do not produce anticorrelation. This result could also be derived from the result of Chap. 6 (Eq. (6.61)):

$$g^{(2)}(0) = 1 + \frac{(\Delta n)^2 - \langle n \rangle}{\langle n \rangle^2} \qquad (11.24)$$

For the coherent state, we found $(\Delta n)^2 = \langle n \rangle$, that is, the shot noise limit. Thus, according to Eq. (11.24), $g^{(2)}(0) = 1$, identical to Eq. (11.23).

Exercise 11.2 What is the probability of detection at D_3 and D_4 if the coherent source in Fig. 11.1 is replaced with a Fock state, $|n\rangle_1$? What would be the probability of double detection? How does this differ from the results for a coherent state? What measurements could you make to distinguish the Fock state from the coherent state?

Chapter 12
Incoherent State

In the previous chapter, we studied coherent states like that produced by a laser. However, the light encountered in almost all situations of everyday life is incoherent light (also called chaotic light or thermal light)—for example, from an incandescent source (resistance filament), the Sun, blackbody radiation, etc. In this chapter, the quantum optics treatment of incoherent or thermal light is introduced. We derive the properties of incoherent light, including the photon number distribution and correlation function, and compare with other types of light.

12.1 Incoherent State

Recall from Chap. 10 that coherent sources are described by a well-defined phase:

$$\alpha = |\alpha|e^{i\varphi} \tag{12.1}$$

In contrast, incoherent (thermal) sources are made of many independent emitters:

$$|\psi\rangle = |\alpha_1\rangle \otimes |\alpha_2\rangle \otimes \ldots |\alpha_l\rangle \ldots = \Pi_l|\alpha_l\rangle \tag{12.2}$$

$$\alpha_l = |\alpha_l|e^{i\varphi_l} \tag{12.3}$$

The incoherent state is described by a multimode state where the phase φ_l is randomly distributed from 0 to 2π.

© The Author(s), under exclusive license to Springer Nature Switzerland AG 2022   115
R. LaPierre, *Getting Started in Quantum Optics*, Undergraduate Texts in Physics,
https://doi.org/10.1007/978-3-031-12432-7_12

12.2   Electric Field of Incoherent State

Before describing the incoherent state, we summarize some key results for the
coherent state. The electric field operator for a mode l is

$$\widehat{\boldsymbol{E}}(\boldsymbol{r}, t) = i\boldsymbol{\varepsilon}_l\varepsilon_l^1 \left(\widehat{a}_l e^{i(\boldsymbol{k}_l\cdot\boldsymbol{r}-\omega_l t)} - \widehat{a}_l^\dagger e^{-i(\boldsymbol{k}_l\cdot\boldsymbol{r}-\omega_l t)} \right) \tag{12.4}$$

where we have included the time-dependence (Heisenberg picture). Equation (12.4)
can be written as a sum of positive and negative frequency components:

$$\widehat{\boldsymbol{E}}(\boldsymbol{r}, t) = \widehat{\boldsymbol{E}}^+(\boldsymbol{r}, t) + \widehat{\boldsymbol{E}}^-(\boldsymbol{r}, t) \tag{12.5}$$

where

$$\widehat{\boldsymbol{E}}^+(\boldsymbol{r}, t) = i\boldsymbol{\varepsilon}_l\varepsilon_l^1\widehat{a}_l e^{i(\boldsymbol{k}_l\cdot\boldsymbol{r}-\omega_l t)} \tag{12.6}$$

and $\widehat{\boldsymbol{E}}^-(\boldsymbol{r}, t)$ is the Hermitian conjugate of $\widehat{\boldsymbol{E}}^+(\boldsymbol{r}, t)$; that is, $\widehat{\boldsymbol{E}}^-(\boldsymbol{r}, t) = \left[\widehat{\boldsymbol{E}}^+(\boldsymbol{r}, t)\right]^\dagger$.
When applied to the coherent state, Eq. (12.5) gives the average electric field as

$$\langle\boldsymbol{E}(\boldsymbol{r}, t)\rangle = i\boldsymbol{\varepsilon}_l\varepsilon_l^1\alpha_l e^{i(\boldsymbol{k}_l\cdot\boldsymbol{r}-\omega_l t)} + c.c. \tag{12.7}$$

which looks like a classical electric field.

The electric field for an incoherent state can be written as a sum of coherent fields
with random phases:

$$\langle\boldsymbol{E}(\boldsymbol{r}, t)\rangle = \sum_l i\boldsymbol{\varepsilon}_l\varepsilon_l^1|\alpha_l|e^{i\varphi_l}e^{i(\boldsymbol{k}_l\cdot\boldsymbol{r}-\omega_l t)} + c.c. \tag{12.8}$$

which can be written as a sum of positive and negative frequency components:

$$\langle\boldsymbol{E}(\boldsymbol{r}, t)\rangle = \boldsymbol{E}^+(\boldsymbol{r}, t) + \boldsymbol{E}^-(\boldsymbol{r}, t) \tag{12.9}$$

where

$$\boldsymbol{E}^+(\boldsymbol{r}, t) = \sum_l i\boldsymbol{\varepsilon}_l\varepsilon_l^1|\alpha_l|e^{i\varphi_l}e^{i(\boldsymbol{k}_l\cdot\boldsymbol{r}-\omega_l t)} \tag{12.10}$$

and $\boldsymbol{E}^-(\boldsymbol{r}, t)$ is the complex conjugate of $\boldsymbol{E}^+(\boldsymbol{r}, t)$; that is, $\boldsymbol{E}^-(\boldsymbol{r}, t) = [\boldsymbol{E}^+(\boldsymbol{r}, t)]^*$. In the
incoherent state, the phases φ_l of each mode are randomly distributed, so the
summation involving $e^{i\varphi_l}$ is zero. The result is a summation of phasors with random
phase angles, resulting in zero average electric field. Equivalently, the average over
many modes l of $e^{i\varphi_l}$, represented by an overbar, is zero:

$$\overline{e^{i\varphi_l}} = 0 \qquad (12.11)$$

Thus, from Eqs. (12.9) and (12.10), the average electric field is zero:

$$\langle E(r, \ t)\rangle = 0 \qquad (12.12)$$

12.3   Photodetector Signal

Similar to Chap. 5, the photodetector signal (intensity or number of counts) associated with the incoherent state can be written as

$$I \propto \left\| \widehat{E}^{+}(r, t)|\psi\rangle \right\|^{2} \qquad (12.13)$$

where $|\psi\rangle$ is the incoherent state and $\widehat{E}^{+}(r, t) = \sum_l i\varepsilon_l \varepsilon_l^1 \widehat{a} e^{i(k_l \cdot r - \omega_l t)}$ is the positive frequency component of the field operator. $E^{+}(r, t)$ is the eigenvalue of the operator $\widehat{E}^{+}(r, t)$. Thus, Eq. (12.13) becomes

$$I \propto \left\| E^{+}(r, t)|\psi\rangle \right\|^{2} = \left| E^{+}(r, \ t) \right|^{2} = E^{-}(r, t)E^{+}(r, t) \qquad (12.14)$$

where $E^{+}(r, t)$ is given by Eq. (12.10). This results in a double summation:

$$I \propto E^{-}(r, t)E^{+}(r, t) = \Sigma_l \Sigma_m \varepsilon_l^1 \varepsilon_m^1 |\alpha_l||\alpha_m| e^{i(\varphi_l - \varphi_m)} e^{i[(k_l - k_m)\cdot r - (\omega_l - \omega_m)t]} \qquad (12.15)$$

The summation only contributes a finite value when $\varphi_l = \varphi_m$; otherwise, the summation over many modes is zero. Equivalently, the average of the phase term is

$$\overline{e^{i(\varphi_l - \varphi_m)}} = \delta_{lm} \qquad (12.16)$$

Hence, with $l = m$ in Eq. (12.15), we get the measured intensity as

$$I \propto \sum_l \left(\varepsilon_l^1\right)^2 |\alpha_l|^2 \qquad (12.17)$$

which is equal to the sum of the intensities from the individual modes without any cross terms from different modes.

12.4   Photon Number Distribution

Max Planck (Fig. 12.1a) correctly described the blackbody radiation spectrum (Fig. 12.1b) by treating the atomic vibrations of the material at finite temperature as quantum harmonic oscillators (QHOs) [1]. Later, Einstein proposed the quantization of light itself. The probability of a QHO having energy $E_n = \hbar\omega\left(n + \frac{1}{2}\right)$ is given by a Boltzmann distribution:

$$P_n = \frac{e^{-E_n/kT}}{Z} \tag{12.18}$$

where Z is the partition function familiar from statistical thermodynamics:

$$Z = \sum_{n=0}^{\infty} e^{-E_n/kT} \tag{12.19}$$

Z is a normalization factor that ensures $\sum_{n=0}^{\infty} P_n = 1$. Thus, substituting $E_n = \hbar\omega\left(n + \frac{1}{2}\right)$ into Eq. (12.19) gives

$$Z = e^{-\hbar\omega/2kT} \sum_{n=0}^{\infty} e^{-n\hbar\omega/kT} \tag{12.20}$$

Eq. (12.20) contains a geometric series:

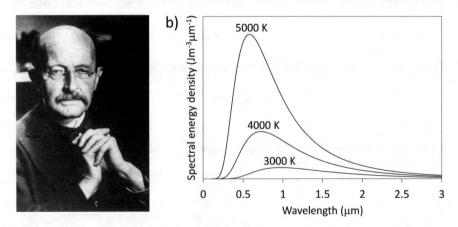

Fig. 12.1 (a) Max Planck (Nobel Prize in Physics in 1918). (Credit: AIP Emilio Segrè Visual Archives, Gift of Jost Lemmerich). (b) Blackbody radiation spectrum for different temperatures

$$\sum_{n=0}^{\infty} e^{-n\hbar\omega/kT} = \frac{1}{1 - e^{-\hbar\omega/kT}} \tag{12.21}$$

which gives

$$Z = \frac{e^{-\hbar\omega/2kT}}{1 - e^{-\hbar\omega/kT}} \tag{12.22}$$

Hence,

$$P_n = e^{-n\hbar\omega/kT}\left(1 - e^{-\hbar\omega/kT}\right) \tag{12.23}$$

P_n ultimately leads to the Planck distribution law for blackbody radiation derived in Appendix 4 and shown in Fig. 12.1b.

From Eq. (12.23), the average photon number is (Exercise 12.1)

$$\langle n \rangle = \sum_{n=0}^{\infty} nP_n = \frac{1}{e^{\hbar\omega/kT} - 1} \tag{12.24}$$

Equation (12.24) is the famous Bose–Einstein distribution for photons. For example, at room temperature (300 K) and $\lambda=500$ nm, we get $\langle n \rangle \sim 10^{-42}$! At 6000 K (surface of the Sun) and $\lambda=500$ nm, we still only get $\langle n \rangle \sim 10^{-2}$.

Exercise 12.1 Derive Eq. (12.24).

From Eqs. (12.23) and (12.24), we get (Exercise 12.2):

$$P_n = \frac{\langle n \rangle^n}{(1 + \langle n \rangle)^{n+1}} \tag{12.25}$$

This probability distribution is shown in Fig. 12.2 for various values of $\langle n \rangle$. Note that the most probable photon number is $n = 0$, and the probability distribution decays exponentially with n as expected from Eq. (12.18).

Exercise 12.2 Derive Eq. (12.25).

12.5 Photon Number Uncertainty

To calculate the photon number uncertainty, Δn, for incoherent light, we start with the following relation:

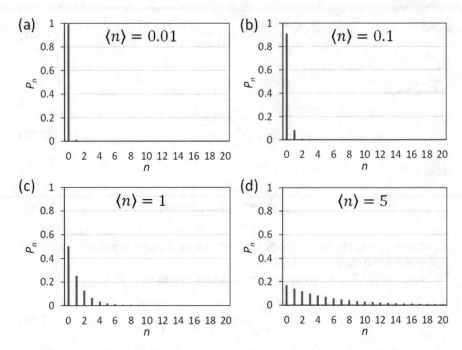

Fig. 12.2 Probability distribution P_n for thermal light with (**a**) $\langle n \rangle = 0.01$, (**b**) $\langle n \rangle = 0.1$, (**c**) $\langle n \rangle = 1$, and (**d**) $\langle n \rangle = 5$

$$\langle n^2 \rangle - \langle n \rangle = \langle n \rangle (\langle n \rangle - 1) = \sum_n n(n-1) P_n \tag{12.26}$$

Evaluating the summation gives (Exercise 12.3)

$$\sum_n n(n-1) P_n = 2\langle n \rangle^2 \tag{12.27}$$

Thus, from Eq. (12.26),

$$\langle n^2 \rangle = 2\langle n \rangle^2 + \langle n \rangle \tag{12.28}$$

Finally, the uncertainty is

$$\Delta n = \sqrt{\langle n^2 \rangle - \langle n \rangle^2} \tag{12.29}$$

Substituting Eq. (12.28) gives

$$\Delta n = \sqrt{\langle n \rangle^2 + \langle n \rangle} \tag{12.30}$$

$$= \sqrt{\langle n \rangle (\langle n \rangle + 1)} \tag{12.31}$$

The uncertainty or fluctuation described by Eq. (12.31) for incoherent light is super-Poissonian; that is, the fluctuations are greater than Poissonian where Δn was equal to $\sqrt{\langle n \rangle}$ (the shot noise limit).

Exercise 12.3 Derive Eq. (12.27).

12.6   Comparison of Different Types of Light

Now that we have described incoherent light, we can review the different types of light covered thus far. A comparison of the photon number distribution for the different types of light is shown in Fig. 12.3 for an average photon number $\langle n \rangle = 5$. The Fock state, $|5\rangle$, has 5 photons with no uncertainty ($\Delta n = 0$). The coherent state follows a Poisson distribution, while the incoherent light (thermal light) has the distribution given by Eq. (12.25). As might be expected, the thermal distribution is broader than that from a coherent source (note the change in horizontal scale).

Exercise 12.4 Plot the probability distributions in Fig. 12.3 for $\langle n \rangle = 1$.

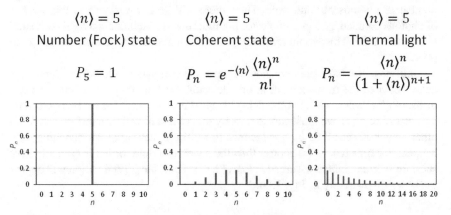

Fig. 12.3 Comparison of photon number distribution for different types of light for $\langle n \rangle = 5$

Table 12.1 Correlation function for different types of light

State	$g^2(0)$	Photon statistics
Classical	≥ 1	Super-Poissonian, Bunched
Fock state	<1	Sub-Poissonian, Antibunched
Coherent state	1	Poissonian
Incoherent state	2	Super-Poissonian, Bunched

Recall that we can measure correlations or coincidences between two detectors in a Hanbury Brown–Twiss experiment and obtain the correlation function $g^2(0)$, which is an important method of characterizing light sources. The second-order correlation function, $g^2(0)$, for thermal light can be derived from the result of Chap. 6:

$$g^{(2)}(0) = 1 + \frac{(\Delta n)^2 - \langle n \rangle}{\langle n \rangle^2} \tag{12.32}$$

For thermal light, we found $(\Delta n)^2 = \langle n \rangle^2 + \langle n \rangle$. Thus, according to Eq. (12.32), $g^{(2)}(0) = 2$. Even if you take a thermal source and spectrally and spatially filter it to look like a laser, it can be distinguished from a laser by a second-order correlation function of 2 rather than 1 for a laser (coherent source). The thermal state has a higher probability to emit more than one photon at the same time. This effect is called photon bunching. Table 12.1 summarizes the correlation function for different types of light.

Recall that the variance of the photon number for a Fock state is zero. Any light with a sub-Poissonian distribution of photon number, like the Fock state, is called antibunched. Antibunched photons are distributed more uniformly in time as compared to the photons in a coherent (Poissonian) beam having the same average number of photons per unit time. Thus, there will be less variance in the number of photons counted by a photodetector per unit time for antibunched light compared to coherent light. This would reduce any noise in a measurement based on counting photons.

Until now, we have only considered the $g^2(0)$ correlation function with no time delay between the measurement at one detector and the other. In general, we can measure $g^2(\tau)$ where τ is the time delay between measurements at the two detectors. To determine $g^2(\tau)$, we must consider many modes and how they interfere with each other, which is described by a coherence time. The bunching or antibunching only happens for time delays (τ) shorter than the coherence time, which is typically very short for thermal light as assumed in Eq. (12.17). Thus, the $g^2(\tau)$ function will appear as shown qualitatively in Fig. 12.4.

Fig. 12.4 Correlation function, $g^2(\tau)$, for different types of light: thermal or bunched light (red), coherent light (green), and antibunched light from a Fock or number state (the case of a single photon source (SPS) is shown; blue)

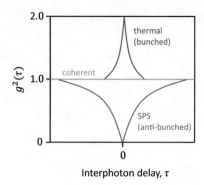

Reference

1. M. Planck, Annalen der Physik 4 (1901) 553-563.

Chapter 13
Homodyne and Heterodyne Detection

In this chapter, homodyne detection and heterodyne detection on a beam splitter are introduced. Homodyne detection is a powerful yet simple method for measuring the quadrature components of electric field. Using heterodyne detection, we can down-convert a high-frequency signal to lower frequency.

13.1 Homodyne Detection

In this section, we consider the method of homodyne detection (Fig. 13.1). Homodyne detection means that the frequency of the state on each of the two input ports of a beam splitter or interferometer is equal:

$$\text{Homodyne detection: } \omega_1 = \omega_2 \qquad (13.1)$$

Consider a beam splitter with reflection coefficient and transmission coefficient equal to $\frac{1}{\sqrt{2}}$:

$$r = t = \frac{1}{\sqrt{2}} \qquad (13.2)$$

This gives a reflectance $R = \frac{1}{2}$ and transmittance $T = \frac{1}{2}$, that is, a 50:50 beam splitter. The input state is

$$|\psi_{\text{in}}\rangle = |\psi\rangle_1 |\alpha_{\text{LO}}\rangle_2 \qquad (13.3)$$

where $|\alpha_{\text{LO}}\rangle_2$ is a coherent state called the "local oscillator" (LO), and $|\psi\rangle_1$ is also typically a coherent state.

Fig. 13.1 Homodyne
detection using a beam
splitter

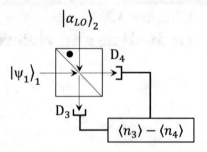

The homodyne signal is the difference in photodetector signals measured between
detectors D_3 and D_4, which is proportional to the difference in photon number
counts, $\langle n_3 \rangle - \langle n_4 \rangle$:

$$\text{Homodyne signal} \propto \langle n_3 \rangle - \langle n_4 \rangle \tag{13.4}$$

Let us begin by calculating $\langle n_3 \rangle$:

$$\langle n_3 \rangle = \langle \psi_{\text{out}} | \hat{a}_3^\dagger \hat{a}_3 | \psi_{\text{out}} \rangle \tag{13.5}$$

Expressed in terms of the input state, we have

$$\langle n_3 \rangle = \langle \psi_{\text{in}} | \frac{1}{2} (\hat{a}_1^\dagger + \hat{a}_2^\dagger)(\hat{a}_1 + \hat{a}_2) | \psi_{\text{in}} \rangle \tag{13.6}$$

$$= \langle \psi_{\text{in}} | \frac{1}{2} (\hat{a}_1^\dagger \hat{a}_1 + \hat{a}_1^\dagger \hat{a}_2 + \hat{a}_2^\dagger \hat{a}_1 + \hat{a}_2^\dagger \hat{a}_2) | \psi_{\text{in}} \rangle \tag{13.7}$$

The factor of $\frac{1}{2}$ comes from the reflection and transmission coefficients. Similarly,

$$\langle n_4 \rangle = \langle \psi_{\text{out}} | \hat{a}_4^\dagger \hat{a}_4 | \psi_{\text{out}} \rangle \tag{13.8}$$

$$= \langle \psi_{\text{in}} | \frac{1}{2} (\hat{a}_1^\dagger - \hat{a}_2^\dagger)(\hat{a}_1 - \hat{a}_2) | \psi_{\text{in}} \rangle \tag{13.9}$$

$$= \langle \psi_{\text{in}} | \frac{1}{2} \left(\hat{a}_1^\dagger \hat{a}_1 - \hat{a}_1^\dagger \hat{a}_2 - \hat{a}_2^\dagger \hat{a}_1 + \hat{a}_2^\dagger \hat{a}_2 \right) | \psi_{\text{in}} \rangle \tag{13.10}$$

Thus, the homodyne signal is

$$\langle n_3 \rangle - \langle n_4 \rangle = \langle \psi_{\text{in}} | \left(\hat{a}_1^\dagger \hat{a}_2 + \hat{a}_2^\dagger \hat{a}_1 \right) | \psi_{\text{in}} \rangle \tag{13.11}$$

$$= {}_2\langle \alpha_{\text{LO}} |_1 \langle \psi_1 | \left(\hat{a}_1^\dagger \hat{a}_2 + \hat{a}_2^\dagger \hat{a}_1 \right) | \psi_1 \rangle_1 | \alpha_{\text{LO}} \rangle_2 \tag{13.12}$$

$$= \langle \psi_1 | (\alpha_{LO}\hat{a}_1^\dagger + \alpha_{LO}^*\hat{a}_1) | \psi_1 \rangle \qquad (13.13)$$

Note that $\hat{n}_3 - \hat{n}_4$ is Hermitian, and therefore can be an observable.

In general, α_{LO} is a complex number and can be written as

$$\alpha_{LO} = |\alpha_{LO}|e^{i\varphi_{LO}} \qquad (13.14)$$

Hence,

$$\langle n_3 \rangle - \langle n_4 \rangle = |\alpha_{LO}| \langle \psi_1 | \left(e^{i\varphi_{LO}}\hat{a}_1^\dagger + e^{-i\varphi_{LO}}\hat{a}_1 \right) | \psi_1 \rangle \qquad (13.15)$$

Using the Euler relation for $e^{i\varphi_{LO}}$ and $e^{-i\varphi_{LO}}$, we get

$$\langle n_3 \rangle - \langle n_4 \rangle = |\alpha_{LO}|$$
$$\times \left\langle \psi_1 \left| \left[(\cos\varphi_{LO} + i\sin\varphi_{LO})\hat{a}_1^\dagger + (\cos\varphi_{LO} - i\sin\varphi_{LO})\hat{a}_1 \right] \right| \psi_1 \right\rangle \qquad (13.16)$$

Rearranging, we obtain

$$\langle n_3 \rangle - \langle n_4 \rangle = \sqrt{2}|\alpha_{LO}|[\cos\varphi_{LO}\langle\psi_1|\frac{1}{\sqrt{2}}\left(\hat{a}_1^\dagger + \hat{a}_1\right)|\psi_1\rangle \qquad (13.17)$$
$$+ \sin\varphi_{LO}\langle\psi_1|\frac{i}{\sqrt{2}}(\hat{a}_1^\dagger - \hat{a}_1)|\psi_1\rangle]$$

Finally, from the definition of the quadrature operators $\hat{Q}$ and $\hat{P}$ in Eqs. (2.44) and (2.45), respectively, we obtain

$$\langle n_3 \rangle - \langle n_4 \rangle = \sqrt{2}|\alpha_{LO}|[\cos\varphi_{LO}\langle\psi_1|\hat{Q}|\psi_1\rangle + \sin\varphi_{LO}\langle\psi_1|\hat{P}|\psi_1\rangle] \qquad (13.18)$$

Thus, we can measure each quadrature of the state $|\psi_1\rangle$ by choosing the phase of the local oscillator, $\varphi_{LO} = 0$ or $\pi/2$. Also, note that $|\alpha_{LO}|$ can be large, giving amplification to the homodyne signal.

Suppose $|\psi_1\rangle$ is a coherent state, $|\psi_1\rangle = |\alpha_1\rangle$, with eigenvalue given by

$$\alpha_1 = |\alpha_1|e^{i\varphi_1} \qquad (13.19)$$

Then,

Fig. 13.2 Quadrature
representation of a coherent
state. The Q quadrature is
obtained when $\varphi_{LO} = 0$, and
the P quadrature is obtained
when $\varphi_{LO} = \pi/2$

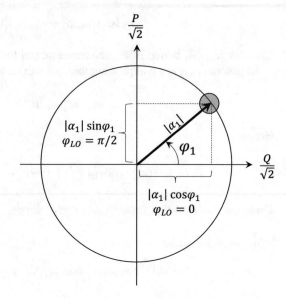

$$Q = \langle Q \rangle = \langle \alpha_1 | \widehat{Q} | \alpha_1 \rangle = \frac{1}{\sqrt{2}}(\alpha_1^* + \alpha_1) = \sqrt{2}\,\mathrm{Re}(\alpha_1) = \sqrt{2}|\alpha_1| \cos \varphi_1 \quad (13.20)$$

$$P = \langle P \rangle = \langle \alpha_1 | \widehat{P} | \alpha_1 \rangle = \frac{i}{\sqrt{2}}(\alpha_1^* - \alpha_1) = \sqrt{2}\mathrm{Im}(\alpha_1) = \sqrt{2}|\alpha_1| \sin \varphi_1 \quad (13.21)$$

and

$$\langle n_3 \rangle - \langle n_4 \rangle = 2|\alpha_{LO}| \left[\cos \varphi_{LO}(|\alpha_1| \cos \varphi_1) + \sin \varphi_{LO}(|\alpha_1| \sin \varphi_1) \right] \quad (13.22)$$

We see that Q and P are related to the real and imaginary parts of α_1, and can be represented in the complex plane as shown in Fig. 13.2. We already saw this in Chap. 10, but now we have a way of measuring the quadrature components by selection of the local oscillator phase φ_{LO} using homodyne detection. The Q quadrature is obtained when $\varphi_{LO} = 0$, and the P quadrature is obtained when $\varphi_{LO} = \pi/2$. We can think of the local oscillator as a "strobe light" that takes a snapshot of the light field of $|\alpha_1\rangle$ at periodic times.

13.2 Heterodyne Detection

In the previous section, we considered homodyne detection where the frequency of the two input signals is equal. In this section, we briefly consider heterodyne detection, where the two input ports contain coherent sources, $|\alpha_1\rangle_1$ and $|\alpha_2\rangle_2$, at

Fig. 13.3 Heterodyne detection using a beam splitter

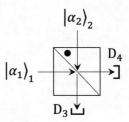

two different frequencies, $\omega_1 \neq \omega_2$, as shown in Fig. 13.3. $|\alpha_2\rangle_2$ is the local oscillator, while $|\alpha_1\rangle_1$ is the signal of interest.

Let us calculate the average photon number at D_3:

$$\langle n_3 \rangle = \langle \psi_{\text{out}} | \hat{a}_3^\dagger \hat{a}_3 | \psi_{\text{out}} \rangle \tag{13.23}$$

$$= \langle \psi_{\text{in}} | \frac{1}{2} (\hat{a}_1^\dagger + \hat{a}_2^\dagger)(\hat{a}_1 + \hat{a}_2) | \psi_{\text{in}} \rangle \tag{13.24}$$

$$= \langle \psi_{\text{in}} | \frac{1}{2} \left(\hat{a}_1^\dagger \hat{a}_1 + \hat{a}_1^\dagger \hat{a}_2 + \hat{a}_2^\dagger \hat{a}_1 + \hat{a}_2^\dagger \hat{a}_2 \right) | \psi_{\text{in}} \rangle \tag{13.25}$$

$$= \frac{1}{2} {}_2\langle \alpha_2 |_1 \langle \alpha_1 | \left(\hat{a}_1^\dagger \hat{a}_1 + \hat{a}_1^\dagger \hat{a}_2 + \hat{a}_2^\dagger \hat{a}_1 + \hat{a}_2^\dagger \hat{a}_2 \right) | \alpha_1 \rangle_1 | \alpha_2 \rangle_2 \tag{13.26}$$

If we include the time-dependence in the operators (Heisenberg picture), we get

$$\langle n_3(t) \rangle = \frac{1}{2} \left(|\alpha_1|^2 + |\alpha_2|^2 + |\alpha_1||\alpha_2| e^{i[(\omega_1 - \omega_2)t + \varphi_2 - \varphi_1]} + |\alpha_1||\alpha_2| e^{-i[(\omega_1 - \omega_2)t + \varphi_2 - \varphi_1]} \right) \tag{13.27}$$

$$= \frac{1}{2} \left(|\alpha_1|^2 + |\alpha_2|^2 \right) + |\alpha_1||\alpha_2| \cos \left[(\omega_1 - \omega_2)t + \varphi_2 - \varphi_1 \right] \tag{13.28}$$

We obtain the familiar form for the interference of two coherent signals. A heterodyne signal with angular frequency, $\omega_1 - \omega_2$, is observed. Using heterodyne detection, small-frequency shifts of a signal from the local oscillator frequency can be measured, which is important for many applications such as Doppler lidar.

Exercise 13.1 Describe some other applications of heterodyne detection.

Chapter 14
Coherent State in an Interferometer

We revisit the coherent state and determine the probability of single and double photon detection in an interferometer and the expression for homodyne detection. The uncertainty and signal-to-noise ratio (SNR) of the homodyne signal is analyzed, leading to an important conclusion—the SNR for the coherent state arises from the uncertainty in the field quadrature of the vacuum input to the interferometer.

14.1 Coherent Light Interference

Let us examine the output of a coherent state in an interferometer as shown in Fig. 14.1. We have

$$\hat{a}_3 = \left(t^2 e^{ikz_1} - r^2 e^{ikz_2}\right)\hat{a}_1 + \left(-rt e^{ikz_1} - tr e^{ikz_2}\right)\hat{a}_2 \tag{14.1}$$

The probability of detection at D_3 is

$$P_3 = \frac{\langle n_3 \rangle}{\langle n_{in} \rangle} = \frac{\langle \Psi_{out}|\hat{n}_3|\Psi_{out}\rangle}{|\alpha|^2} \tag{14.2}$$

Converting to the input state, retaining only the non-zero terms, and assuming r and t are real numbers gives

$$P_3 = \frac{{}_2\langle 0|_1\langle \alpha| \left[\left(t^2 e^{-ikz_1} - r^2 e^{-ikz_2}\right)\left(t^2 e^{ikz_1} - r^2 e^{ikz_2}\right)\hat{a}_1^\dagger \hat{a}_1\right]|\alpha\rangle_1|0\rangle_2}{|\alpha|^2} \tag{14.3}$$

$$= \left[R^2 + T^2 - 2RT\cos(k\Delta z)\right] \tag{14.4}$$

© The Author(s), under exclusive license to Springer Nature Switzerland AG 2022 131
R. LaPierre, *Getting Started in Quantum Optics*, Undergraduate Texts in Physics,
https://doi.org/10.1007/978-3-031-12432-7_14

Fig. 14.1 Coherent state in
an interferometer

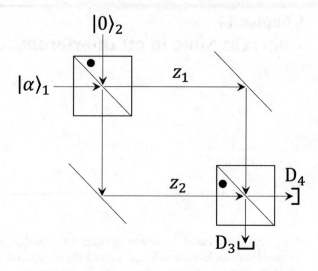

which is the same as the classical result. For the usual case of a 50:50 beam splitter,
Eq. (14.4) gives

$$P_3 = \sin^2\left(\frac{k\Delta z}{2}\right) \tag{14.5}$$

Similarly,

$$\hat{a}_4 = \left(rte^{ikz_1} + tre^{ikz_2}\right)\hat{a}_1 + \left(-r^2e^{ikz_1} + t^2e^{ikz_2}\right)\hat{a}_2 \tag{14.6}$$

The probability of detection at D_4 is

$$P_4 = \frac{\langle n_4\rangle}{\langle n_{\text{in}}\rangle} = \frac{\langle\psi_{\text{out}}|\hat{n}_4|\psi_{\text{out}}\rangle}{|\alpha|^2} \tag{14.7}$$

Converting to the input state, retaining only the non-zero terms, and assuming r and
t are real numbers gives

$$P_4 = \frac{{}_2\langle 0|{}_1\langle\alpha|\left[\left(rte^{-ikz_1} + tre^{-ikz_2}\right)\left(rte^{ikz_1} + tre^{ikz_2}\right)\hat{a}_1^\dagger\hat{a}_1\right]|\alpha\rangle_1|0\rangle_2}{|\alpha|^2} \tag{14.8}$$

$$= 2RT + 2RT\cos\left(k\Delta z\right) \tag{14.9}$$

which is the same as the classical result. For the usual case of a 50:50 beam splitter,
Eq. (14.9) gives

$$P_4 = \cos^2\left(\frac{k\Delta z}{2}\right) \tag{14.10}$$

Note that $P_3 + P_4 = 1$, as required for probabilities.

14.2 Coincident Detection

The probability of double detection at D_3 and D_4 is

$$P_{34} = \frac{\langle\Psi_{\text{out}}|\hat{a}_3^\dagger\hat{a}_4^\dagger\hat{a}_4\hat{a}_3|\Psi_{\text{out}}\rangle}{|\alpha|^4} \tag{14.11}$$

Retaining the non-zero terms gives

$$P_{34} = \frac{{}_2\langle 0|{}_1\langle\alpha|\left(t^2 e^{-ikz_1} - r^2 e^{-ikz_2}\right)\left(rte^{-ikz_1} + tre^{-ikz_2}\right)\left(rte^{ikz_1} + tre^{ikz_2}\right)}{|\alpha|^4} \tag{14.12}$$

$$\left(t^2 e^{ikz_1} - r^2 e^{ikz_2}\right)\hat{a}_1^\dagger\hat{a}_1^\dagger\hat{a}_1\hat{a}_1|\alpha\rangle_1|0\rangle_2$$

$$= P_3 P_4 \tag{14.13}$$

which is the same as the classical result.

Exercise 14.1 Derive Eq. (14.13).

14.3 Homodyne Signal

Let us determine the homodyne signal of a coherent state in an interferometer, as shown in Fig. 14.2. Note that we moved the coherent state to port 2. The input state is

$$|\Psi_{\text{in}}\rangle = |0\rangle_1|\alpha\rangle_2 \tag{14.14}$$

with vacuum on port 1 and the coherent state on port 2. The operators are

$$\hat{a}_3 = \left(t^2 e^{ikz_1} - r^2 e^{ikz_2}\right)\hat{a}_1 + \left(-rte^{ikz_1} - tre^{ikz_2}\right)\hat{a}_2 \tag{14.15}$$

$$\hat{a}_4 = \left(rte^{ikz_1} + tre^{ikz_2}\right)\hat{a}_1 + \left(-r^2 e^{ikz_1} + t^2 e^{ikz_2}\right)\hat{a}_2 \tag{14.16}$$

Let us consider the usual case of a 50:50 beam splitter. Equations (14.15) and (14.16) simplify to

Fig. 14.2 Homodyne
detection in an
interferometer

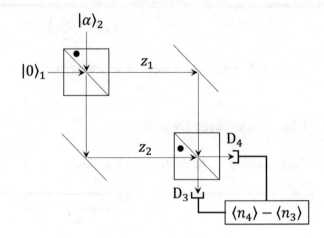

$$\widehat{a}_3 = \frac{1}{2}\left(e^{ikz_1} - e^{ikz_2}\right)\widehat{a}_1 - \frac{1}{2}\left(e^{ikz_1} + e^{ikz_2}\right)\widehat{a}_2 \tag{14.17}$$

$$\widehat{a}_4 = \frac{1}{2}\left(e^{ikz_1} + e^{ikz_2}\right)\widehat{a}_1 - \frac{1}{2}\left(e^{ikz_1} - e^{ikz_2}\right)\widehat{a}_2 \tag{14.18}$$

The annihilation operators are

$$\widehat{a}_3^\dagger = \frac{1}{2}\left(e^{-ikz_1} - e^{-ikz_2}\right)\widehat{a}_1^\dagger - \frac{1}{2}\left(e^{-ikz_1} + e^{-ikz_2}\right)\widehat{a}_2^\dagger \tag{14.19}$$

$$\widehat{a}_4^\dagger = \frac{1}{2}\left(e^{-ikz_1} + e^{-ikz_2}\right)\widehat{a}_1^\dagger - \frac{1}{2}\left(e^{-ikz_1} - e^{-ikz_2}\right)\widehat{a}_2^\dagger \tag{14.20}$$

From Eqs. (14.17), (14.18), (14.19) and (14.20), the number operators are

$$\begin{aligned}
\widehat{n}_3 = \widehat{a}_3^\dagger \widehat{a}_3 = {}& \sin^2(k\Delta z/2)\widehat{a}_1^\dagger\widehat{a}_1 + \frac{i}{2}\sin(k\Delta z)\widehat{a}_1^\dagger\widehat{a}_2 \\
& - \frac{i}{2}\sin(k\Delta z)\widehat{a}_2^\dagger\widehat{a}_1 + \cos^2(k\Delta z/2)\widehat{a}_2^\dagger\widehat{a}_2
\end{aligned} \tag{14.21}$$

and

$$\begin{aligned}
\widehat{n}_4 = \widehat{a}_4^\dagger \widehat{a}_4 = {}& \cos^2(k\Delta z/2)\widehat{a}_1^\dagger\widehat{a}_1 - \frac{i}{2}\sin(k\Delta z)\widehat{a}_1^\dagger\widehat{a}_2 \\
& + \frac{i}{2}\sin(k\Delta z)\widehat{a}_2^\dagger\widehat{a}_1 + \sin^2(k\Delta z/2)\widehat{a}_2^\dagger\widehat{a}_2
\end{aligned} \tag{14.22}$$

where $\Delta z = z_1 - z_2$.

Exercise 14.2 Derive Eqs. (14.21) and (14.22).

The average photon numbers at D_3 and D_4 are

$$\langle n_3 \rangle = {}_2\langle \alpha|_1 \langle 0|\hat{a}_3^\dagger \hat{a}_3|0\rangle_1 |\alpha\rangle_2 = |\alpha|^2 \cos^2(k\Delta z/2) \tag{14.23}$$

$$\langle n_4 \rangle = {}_2\langle \alpha|_1 \langle 0|\hat{a}_4^\dagger \hat{a}_4|0\rangle_1 |\alpha\rangle_2 = |\alpha|^2 \sin^2(k\Delta z/2) \tag{14.24}$$

Exercise 14.3   Derive Eqs. (14.23) and (14.24).

Let us take the homodyne signal as $\langle n_4 - n_3 \rangle = \langle n_4 \rangle - \langle n_3 \rangle$. From Eqs. (14.23) and (14.24), we get

$$\langle n_4 \rangle - \langle n_3 \rangle = |\alpha|^2 \left[\sin^2 \left(\frac{k\Delta z}{2} \right) - \cos^2 \left(\frac{k\Delta z}{2} \right) \right] \tag{14.25}$$

$$= -|\alpha|^2 \cos (k\Delta z) \tag{14.26}$$

This signal is shown in Fig. 14.3.

Suppose we want to measure small displacements (e.g., movements of the mirrors) corresponding to a path length difference or phase change, ε. The greatest sensitivity to displacement occurs near $\varphi = \frac{\pi}{2}$ or $\frac{3\pi}{2}$ where the slope of the homodyne signal, $\langle n_4 \rangle - \langle n_3 \rangle$, in Fig. 14.3 is steepest. Let us consider small phase changes ε from $\frac{\pi}{2}$; that is,

$$k\Delta z = \frac{\pi}{2} + \varepsilon \tag{14.27}$$

Equation (14.26) becomes

$$\langle n_4 \rangle - \langle n_3 \rangle = -|\alpha|^2 \cos \left(\frac{\pi}{2} + \varepsilon \right) \tag{14.28}$$

Fig. 14.3 Homodyne signal for the interferometer in Fig. 14.2

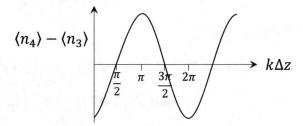

Using a trigonometric identity, Eq. (14.28) becomes

$$\langle n_4 \rangle - \langle n_3 \rangle = -|\alpha|^2 \left[\cos\left(\frac{\pi}{2}\right) \cos(\varepsilon) - \sin\left(\frac{\pi}{2}\right) \sin(\varepsilon) \right] \tag{14.29}$$

$$= |\alpha|^2 \sin \varepsilon \tag{14.30}$$

Since $|\alpha|^2 = \langle n_2 \rangle$ for the coherent state,

$$\langle n_4 \rangle - \langle n_3 \rangle = \langle n_2 \rangle \sin \varepsilon \tag{14.31}$$

According to the Taylor expansion, small ε gives:

$$\langle n_4 \rangle - \langle n_3 \rangle \sim \langle n_2 \rangle \varepsilon \tag{14.32}$$

Equation (14.32) indicates that we have good sensitivity to displacements in the interferometer if $\langle n_2 \rangle$ is large; that is, we want an intense laser.

14.4 Uncertainty in the Homodyne Signal

Let us determine the uncertainty in $\langle n_4 \rangle - \langle n_3 \rangle$. First, we need to determine $\langle (n_4 - n_3)^2 \rangle$. From Eqs. (14.21) and (14.22), we have

$$\hat{n}_4 - \hat{n}_3 = \hat{a}_4^\dagger \hat{a}_4 - \hat{a}_3^\dagger \hat{a}_3 \tag{14.33}$$

$$= \left[\cos^2\left(\frac{k\Delta z}{2}\right) - \sin^2\left(\frac{k\Delta z}{2}\right) \right] \hat{a}_1^\dagger \hat{a}_1 - i \sin(k\Delta z) \hat{a}_1^\dagger \hat{a}_2$$

$$+ i \sin(k\Delta z) \hat{a}_2^\dagger \hat{a}_1 + \left[\sin^2\left(\frac{k\Delta z}{2}\right) - \cos^2\left(\frac{k\Delta z}{2}\right) \right] \hat{a}_2^\dagger \hat{a}_2 \tag{14.34}$$

$$= \cos(k\Delta z)\hat{a}_1^\dagger \hat{a}_1 - i \sin(k\Delta z)\hat{a}_1^\dagger \hat{a}_2 + i \sin(k\Delta z)\hat{a}_2^\dagger \hat{a}_1 - \cos(k\Delta z)\hat{a}_2^\dagger \hat{a}_2 \tag{14.35}$$

$$= \cos(k\Delta z)\left(\hat{a}_1^\dagger \hat{a}_1 - \hat{a}_2^\dagger \hat{a}_2\right) - i \sin(k\Delta z)\left(\hat{a}_1^\dagger \hat{a}_2 - \hat{a}_2^\dagger \hat{a}_1\right) \tag{14.36}$$

Using Eq. (14.27):

$$\hat{n}_4 - \hat{n}_3 = \cos\left(\frac{\pi}{2} + \varepsilon\right)\left(\hat{a}_1^\dagger \hat{a}_1 - \hat{a}_2^\dagger \hat{a}_2\right) - i \sin\left(\frac{\pi}{2} + \varepsilon\right)\left(\hat{a}_1^\dagger \hat{a}_2 - \hat{a}_2^\dagger \hat{a}_1\right) \tag{14.37}$$

$$= -\sin\varepsilon\left(\hat{a}_1^\dagger \hat{a}_1 - \hat{a}_2^\dagger \hat{a}_2\right) - i \cos\varepsilon\left(\hat{a}_1^\dagger \hat{a}_2 - \hat{a}_2^\dagger \hat{a}_1\right) \tag{14.38}$$

Using Eq. (14.38) for small displacements ($\varepsilon \to 0$), $(\hat{n}_4 - \hat{n}_3)^2$ becomes:

$$(\hat{n}_4 - \hat{n}_3)^2 = -\left(\hat{a}_1^\dagger \hat{a}_2 - \hat{a}_2^\dagger \hat{a}_1\right)\left(\hat{a}_1^\dagger \hat{a}_2 - \hat{a}_2^\dagger \hat{a}_1\right) \tag{14.39}$$

$$= -\hat{a}_1^\dagger \hat{a}_2 \hat{a}_1^\dagger \hat{a}_2 + \hat{a}_1^\dagger \hat{a}_2 \hat{a}_2^\dagger \hat{a}_1 + \hat{a}_2^\dagger \hat{a}_1 \hat{a}_1^\dagger \hat{a}_2 - \hat{a}_2^\dagger \hat{a}_1 \hat{a}_2^\dagger \hat{a}_1 \tag{14.40}$$

Using the commutation relation to put the second term in the normal order, we get

$$(\hat{n}_4 - \hat{n}_3)^2 = -\hat{a}_1^\dagger \hat{a}_2 \hat{a}_1^\dagger \hat{a}_2 + \hat{a}_1^\dagger \left(1 + \hat{a}_2^\dagger \hat{a}_2\right)\hat{a}_1 + \hat{a}_2^\dagger \hat{a}_1 \hat{a}_1^\dagger \hat{a}_2 - \hat{a}_2^\dagger \hat{a}_1 \hat{a}_2^\dagger \hat{a}_1 \tag{14.41}$$

Using Eq. (14.41), we can now evaluate the average, $\langle (n_4 - n_3)^2 \rangle$. For simplicity, let us assume α is a real number ($\alpha \in \mathbb{R}$; equivalent to zero phase angle), giving:

$$\langle (n_4 - n_3)^2 \rangle = {}_2\langle \alpha|_1\langle 0| \left[-\alpha^2 \hat{a}_1^\dagger \hat{a}_1^\dagger + (1 + \alpha^2)\hat{a}_1^\dagger \hat{a}_1 + \alpha^2 \hat{a}_1 \hat{a}_1^\dagger - \alpha^2 \hat{a}_1 \hat{a}_1 \right] |0\rangle_1 |\alpha\rangle_2 \tag{14.42}$$

We assume a strong coherent state, so $1 + \alpha^2 \sim \alpha^2$, giving:

$$\langle (n_4 - n_3)^2 \rangle = -\alpha^2 \,{}_1\langle 0| \left(\hat{a}_1^\dagger \hat{a}_1^\dagger - \hat{a}_1^\dagger \hat{a}_1 - \hat{a}_1 \hat{a}_1^\dagger + \hat{a}_1 \hat{a}_1\right)|0\rangle_1 \tag{14.43}$$

The first, second, and fourth terms are zero. Applying the normal ordering to the third term gives

$$\langle (n_4 - n_3)^2 \rangle = \alpha^2 \,{}_1\langle 0|(1 + \hat{a}_1^\dagger \hat{a}_1)|0\rangle_1 \tag{14.44}$$

$$= \alpha^2 = \langle n_2 \rangle \tag{14.45}$$

Finally, we can evaluate the uncertainty:

$$\Delta(n_4 - n_3) = \sqrt{\langle (n_4 - n_3)^2 \rangle - \langle n_4 - n_3 \rangle} \tag{14.46}$$

Using Eqs. (14.32) and (14.45) gives

$$\Delta(n_4 - n_3) = \sqrt{\langle n_2 \rangle - \varepsilon \langle n_2 \rangle} \sim \sqrt{\langle n_2 \rangle} \tag{14.47}$$

The uncertainty is equal to the shot noise limit, as shown in Fig. 14.4. The signal-to-noise ratio (SNR) may be defined as

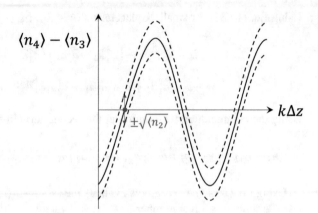

Fig. 14.4 Homodyne signal showing the uncertainty, $\sqrt{\langle n_2 \rangle}$, indicated by the dashed curves and red arrow

$$\text{SNR} = \frac{\langle n_2 \rangle \varepsilon}{\sqrt{\langle n_2 \rangle}} = \sqrt{\langle n_2 \rangle} \, \varepsilon \tag{14.48}$$

The only way to increase the SNR is to increase the source power or the measurement time, corresponding to an increase in $\langle n_2 \rangle$. This approach has its limitations due to damage to the optical system caused by exposure to high optical power.

How does the $\sqrt{\langle n_2 \rangle}$ uncertainty arise? From Eq. (14.43), we have

$$\langle (n_4 - n_3)^2 \rangle = -\alpha^2 \, _1\langle 0| \left(\hat{a}_1^\dagger - \hat{a}_1 \right)^2 |0\rangle_1 \tag{14.49}$$

Recall that the $\widehat{P}$ operator is

$$\widehat{P} = \frac{-i}{\sqrt{2}} \left(\hat{a} - \hat{a}^\dagger \right) \tag{14.50}$$

Thus,

$$\langle (n_4 - n_3)^2 \rangle = 2\alpha^2 \, _1\langle 0| \widehat{P}^2 |0\rangle_1 \tag{14.51}$$

$$= 2\langle n_2 \rangle (\Delta P)^2 \tag{14.52}$$

If we substitute $\Delta P = \frac{1}{\sqrt{2}}$ in Eq. (14.52) for the vacuum state, we reproduce Eq. (14.45). We see that the SNR for the coherent state arises from the uncertainty in the P quadrature of the vacuum on port 1! This is another of the experimental consequences of the vacuum state, in addition to others presented in Chap. 4 (Lamb shift, Casimir effect, etc.). In the next two chapters, we will see how we can use this insight to improve the SNR in an interferometer below the shot noise limit.

Chapter 15
Squeezed Light

Thus far, we have described Fock states, coherent light, and incoherent light. Quantum optics has discovered many other states of light. In this chapter, we examine one of the most useful of these, called "squeezed light". We show that the uncertainty in the phase or amplitude quadrature of squeezed light can be reduced as compared to coherent light, making squeezed light very useful in metrology. The squeezed vacuum state is introduced, and the fragility of the squeezed state is explained.

15.1   Classical Description of Nonlinear Optics

One of the new states of light discovered in quantum optics is "squeezed light", which allows us to surpass the shot noise limit. Although squeezed light had been theoretically studied for a long time, the first experimental success for producing squeezed light was by Slusher et al. in 1985 [1]. The most efficient means of producing squeezed light has used parametric oscillation in nonlinear media. Examples of nonlinear materials include lithium niobate ($LiNbO_3$) or potassium titanyl phosphate (KTP).

The classical description of the parametric oscillator involves the nonlinear dielectric polarization (dipole moment per unit volume). "Parametric" means that some parameter of the oscillator varies periodically in time (e.g., periodically varying the length of a swinging pendulum). In the optical parametric process, a strong coherent source (a pump laser) modulates the polarization of a nonlinear material due to a nonlinear dependence of the polarization P on the applied field. The pump field is a strong coherent field (laser), which is our parametric drive. The dependence of the polarization on the total electric field is given by (do not confuse polarization P with the P quadrature)

© The Author(s), under exclusive license to Springer Nature Switzerland AG 2022
R. LaPierre, *Getting Started in Quantum Optics*, Undergraduate Texts in Physics,
https://doi.org/10.1007/978-3-031-12432-7_15

Fig. 15.1 Production of squeezed light using a nonlinear material

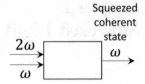

$$P(E) = \underbrace{\epsilon_0 \chi^{(1)} E}_{P^{(1)}} + \underbrace{\epsilon_0 \chi^{(2)} E^2}_{P^{(2)}} + \cdots \text{higher}-\text{order terms} \qquad (15.1)$$

where $\chi^{(1)}$ is the first-order linear electric susceptibility, $\chi^{(2)}$ is the second-order electric susceptibility (a nonlinear term), and so on for higher-order nonlinear terms. Equation (15.1) can be understood as a power series, which arises physically from an anharmonic crystal potential in the nonlinear medium. Typical values for $\chi^{(1)}$ are on the order of unity ($\chi^{(1)}$ is dimensionless), $\chi^{(2)}$ is typically on the order of 10^{-12} m/V, and successive higher-order terms usually decrease quickly (some terms may be zero due to crystal symmetry). Thus, the nonlinear terms of Eq. (15.1) are much less than the linear term and become evident only in the presence of a strong electric field (e.g., from a laser).

Consider Fig. 15.1 where a coherent field at angular frequency ω and pump field at 2ω are put into a nonlinear material such as KTP. The total applied electric field (assuming the position $r = 0$) can be expressed as

$$E = A \cos(\omega t + \varphi) + B \cos(2\omega t) \qquad (15.2)$$

The output field is proportional to the polarization P of the material. According to Eq. (15.1), the linear term is

$$P^{(1)}(E) = \epsilon_0 \chi^{(1)} [A \cos(\omega t + \varphi) + B \cos(2\omega t)] \qquad (15.3)$$

and the second-order term is

$$P^{(2)}(E) = \epsilon_0 \chi^{(2)} \left[A^2 \cos^2(\omega t + \varphi) + B^2 \cos^2(2\omega t) + 2AB \cos(\omega t + \varphi) \cos(2\omega t) \right] \qquad (15.4)$$

Using trigonometric identities, Eq. (15.4) can be expressed purely in cosine terms:

$$P^{(2)}(E) = \epsilon_0 \chi^{(2)} \left\{ \frac{1}{2}A^2[1 + \cos{(2\omega t + 2\varphi)}] + \frac{1}{2}B^2[1 + \cos{(4\omega t)}] \right. \tag{15.5}$$

$$\left. + AB[\cos{(\omega t - \varphi)} + \cos{(3\omega t + \varphi)}] \right\}$$

$P^{(2)}$ contains a constant component and oscillating components at frequency ω, 2ω, 3ω and 4ω. The ω component of the second-order nonlinear term, $P_\omega^{(2)}$, is

$$P_\omega^{(2)} = \epsilon_0 \chi^{(2)} AB \cos{(\omega t - \varphi)} \tag{15.6}$$

From Eq. (15.3), the linear term of the same frequency is

$$P_\omega^{(1)} = \epsilon_0 \chi^{(1)} A \cos{(\omega t + \varphi)} \tag{15.7}$$

The $P_\omega^{(1)}$ and $P_\omega^{(2)}$ terms can constructively interfere if $\varphi = 0$. This constructive interference process is called optical parametric amplification (OPA). On the other hand, destructive interference occurs if $\varphi = \pi/2$. This destructive interference process can also work on the vacuum state, producing the "squeezed vacuum state". Thus, the optical nonlinearity creates correlations between photons that can be exploited to reduce the noise below the shot noise limit.

15.2 Quantum Description of Squeezing

In Fig. 15.1, we refer to the input and output as mode b and mode a, respectively. In the quantum description of squeezing, two photons at frequency ω (the down-converted photons) are created in the output mode a, and one photon at frequency 2ω is destroyed in the input mode b (see Fig. 5.2b in Chap. 5). The reverse process is also possible (it is also required to make $\hat{H}$ Hermitian). The interaction Hamiltonian takes the form

$$\hat{H} = \hbar g \left(\hat{a}\hat{a}\hat{b}^\dagger + \hat{a}^\dagger \hat{a}^\dagger \hat{b} \right) \tag{15.8}$$

$$= \hbar g \left(\hat{a}^2 \hat{b}^\dagger + \hat{a}^{\dagger 2} \hat{b} \right) \tag{15.9}$$

where g is a coupling constant related to the nonlinear susceptibility terms in Eq. (15.1). We assume the mode b input is a strong coherent state (e.g., a laser), which is our parametric drive. Thus, for a coherent state,

$$\hat{b}|\beta\rangle = \beta|\beta\rangle \tag{15.10}$$

where β is given by

$$\beta = |\beta|e^{i\varphi} \tag{15.11}$$

Equation (15.10) states that the coherent state $|\beta\rangle$ is an eigenstate of the annihilation operator with eigenvalue β, which is the definition of the coherent state. We also assume that $|\beta\rangle$ is an eigenstate of the creation operator with eigenvalue β^*:

$$\hat{b}^\dagger|\beta\rangle \sim \beta^*|\beta\rangle \tag{15.12}$$

In other words, we assume that we can replace the annihilation and creation operators with the complex numbers, β and β^*, respectively, which is valid for a strong coherent classical source where adding or removing one photon from the field makes a negligible difference. We can then rewrite the Hamiltonian in Eq. (15.9) as

$$\hat{H} = \hbar g|\beta|\left(\hat{a}^2 e^{-i\varphi} + \hat{a}^{\dagger 2} e^{i\varphi}\right) \tag{15.13}$$

15.3   Squeezing Operator

Using the unitary evolution operator, we define the squeezing operator, $\hat{S}$:

$$\hat{S} = e^{-\frac{i\hat{H}t}{\hbar}} \tag{15.14}$$

where $\hat{H}$ is the Hamiltonian from Eq. (15.13). Thus, we get

$$\hat{S} = e^{-ig|\beta|t\left(\hat{a}^2 e^{-i\varphi} + \hat{a}^{\dagger 2} e^{i\varphi}\right)} \tag{15.15}$$

Choosing $\varphi = -\pi/2$ gives

$$\hat{S} = e^{\frac{R}{2}\left(\hat{a}^2 - \hat{a}^{\dagger 2}\right)} \tag{15.16}$$

where $R = 2g|\beta|t$ is called the squeeze parameter. The factor of 2 in R is present by convention. We can tune R by selecting the phase φ. Note that $\hat{S}$ is unitary since $\hat{H}$ is Hermitian, resulting in

$$\widehat{S}^{-1}(R) = \widehat{S}^{\dagger}(R) = \widehat{S}(-R) \tag{15.17}$$

What does the squeezing operator $\widehat{S}$ do? The squeezing operator applied to a coherent state produces a new state:

$$\widehat{S}|\alpha\rangle = |\alpha, R\rangle \tag{15.18}$$

The new state is called the squeezed state, denoted by $|\alpha, R\rangle$. To characterize this state, let us find the average electric field, quadrature values (Q, P), and uncertainties $(\Delta E, \Delta Q, \Delta P)$ for this new state.

First, we need to find $\langle \alpha, R|\widehat{a}|\alpha, R\rangle = \langle \alpha|\widehat{S}^{\dagger}\widehat{a}\widehat{S}|\alpha\rangle$. For this purpose, we use the Baker–Hausdorff formula:

$$e^{\widehat{B}}\widehat{A}e^{-\widehat{B}} = \widehat{A} + \frac{1}{1!}\left[\widehat{B}, \widehat{A}\right] + \frac{1}{2!}\left[\widehat{B}, \left[\widehat{B}, \widehat{A}\right]\right] + \frac{1}{3!}\left[\widehat{B}, \left[\widehat{B}, \left[\widehat{B}, \widehat{A}\right]\right]\right] + \dots \tag{15.19}$$

The Baker–Hausdorff formula can be proven using the Taylor expansion of the exponential functions:

$$e^{\widehat{B}}\widehat{A}e^{-\widehat{B}} = \left(1 + \widehat{B} + \frac{\widehat{B}^2}{2!} + \dots\right)\widehat{A}\left(1 - \widehat{B} + \frac{\widehat{B}^2}{2!} + \dots\right) \tag{15.20}$$

$$= \widehat{A} + \left(\widehat{B}\widehat{A} - \widehat{A}\widehat{B}\right) + \frac{1}{2!}\left(\widehat{B}^2\widehat{A} + \widehat{A}\widehat{B}^2 - 2\widehat{B}\widehat{A}\widehat{B}\right) + \dots \tag{15.21}$$

Using the definition of the commutation relation, Eq. (15.21) is identical to Eq. (15.19). We consider the case for which $\widehat{A} = \widehat{a}$ and $\widehat{B} = -\frac{R}{2}\left(\widehat{a}^2 - \widehat{a}^{\dagger 2}\right)$. It is easy to show that $\left[\widehat{B}, \widehat{a}\right] = -R\widehat{a}^{\dagger}$ and $\left[\widehat{B}, \widehat{a}^{\dagger}\right] = -R\widehat{a}$. Thus, after using the Baker–Hausdorff formula, we get

$$\widehat{S}^{\dagger}\widehat{a}\widehat{S} = \widehat{a}\left(1 + \frac{R^2}{2!} + \dots\right) - \widehat{a}^{\dagger}\left(R + \frac{R^3}{3!} + \dots\right) \tag{15.22}$$

The terms in brackets are the Taylor expansion for the hyperbolic functions, $\cosh R$ and $\sinh R$:

$$\widehat{S}^{\dagger}\widehat{a}\widehat{S} = \widehat{a}\cosh R - \widehat{a}^{\dagger}\sinh R \tag{15.23}$$

where

$$\cosh R = \frac{e^R + e^{-R}}{2} \tag{15.24}$$

and

$$\sinh R = \frac{e^R - e^{-R}}{2} \qquad (15.25)$$

Similarly,

$$\widehat{S}^\dagger \widehat{a}^\dagger \widehat{S} = \widehat{a}^\dagger \cosh R - \widehat{a} \sinh R \qquad (15.26)$$

With Eqs. (15.23) and (15.26) in our toolbox, we can find the expectation value and uncertainty for the field and its quadratures, as shown in the following sections.

Exercise 15.1 Show that for $\widehat{A} = \widehat{a}$ and $\widehat{B} = -\frac{R}{2}\left(\widehat{a}^2 - \widehat{a}^{\dagger 2}\right)$, $\left[\widehat{B}, \widehat{a}\right] = -R\widehat{a}^\dagger$ and $\left[\widehat{B}, \widehat{a}^\dagger\right] = -R\widehat{a}$.

15.4   Electric Field of Squeezed Light

Recall that the quantum operator for the electric field (Heisenberg picture) is

$$\widehat{E}(r, t) = i\varepsilon\varepsilon^1\left(\widehat{a}e^{i(k\cdot r - \omega t)} - \widehat{a}^\dagger e^{-i(k\cdot r - \omega t)}\right) \qquad (15.27)$$

The average electric field for the squeezed state is

$$\langle E \rangle = \langle \alpha, R | \widehat{E}(r, t) | \alpha, R \rangle \qquad (15.28)$$

Using Eqs. (15.23) and (15.26) gives

$$\langle E \rangle = i\varepsilon\varepsilon^1 (\alpha \cosh R - \alpha^* \sinh R)e^{i(k\cdot r - \omega t)} + c.c. \qquad (15.29)$$

$$= i\varepsilon\varepsilon^1 \alpha' e^{i(k\cdot r - \omega t)} + c.c. \qquad (15.30)$$

where

$$\alpha' = \alpha \cosh R - \alpha^* \sinh R \qquad (15.31)$$

If α is a real number ($\alpha \in \mathbb{R}$), then

$$\langle E \rangle = -2\varepsilon\varepsilon^1 \alpha' \sin(k\cdot r - \omega t) \qquad (15.32)$$

Equation (15.30) or (15.32) resembles a coherent state with amplitude α', similar to Eqs. (10.42) and (10.43), respectively. Thus, the squeezing of a coherent state $|\alpha\rangle$ produces another coherent state $|\alpha, R\rangle$. However, as we will see below, the new coherent state $|\alpha, R\rangle$ (the squeezed state) has some special properties.

If $\alpha \in \mathbb{R}$, then according to Eqs. (15.24), (15.25) and (15.31),

$$\alpha' = \alpha \left(\frac{e^R + e^{-R}}{2} - \frac{e^R - e^{-R}}{2} \right) \tag{15.33}$$

$$= \alpha\, e^{-R} \tag{15.34}$$

For simplicity, let us choose the position $r = 0$. The uncertainty in electric field becomes (Exercise 15.2)

$$\Delta E = \varepsilon^1 \sqrt{e^{2R} \cos^2(-\omega t) + e^{-2R} \sin^2(-\omega t)} \tag{15.35}$$

If the squeezing parameter, $R < 0$, then the minimum field uncertainty occurs when $\omega t = 0, \pi, \ldots$:

$$R < 0 : \Delta E_{\min} = \varepsilon^1 e^R \quad \text{when } \omega t = 0, \pi, \ldots \tag{15.36}$$

Conversely, the maximum field uncertainty occurs when $t = \frac{\pi}{2}, \frac{3\pi}{2}, \ldots$:

$$R < 0 : \Delta E_{\max} = \varepsilon^1 e^{-R} \quad \text{when } \omega t = \frac{\pi}{2}, \frac{3\pi}{2}, \ldots \tag{15.37}$$

We see that the uncertainty depends on time. There are times ($\omega t = 0, \pi, \ldots$) when the uncertainty is less than ε^1 and there are times ($\omega t = \frac{\pi}{2}, \frac{3\pi}{2}, \ldots$) when the uncertainty is greater than ε^1. We can think of the squeezed state as an oscillating Gaussian wavepacket, like the coherent state (Fig. 10.3), but whose width (standard deviation) varies with time.

Conversely, if $R > 0$:

$$R > 0 : \Delta E_{\min} = \varepsilon^1 e^{-R} \quad \text{when } \omega t = \frac{\pi}{2}, \frac{3\pi}{2}, \ldots \tag{15.38}$$

$$R > 0 : \Delta E_{\max} = \varepsilon^1 e^R \quad \text{when } \omega t = 0, \pi, \ldots \tag{15.39}$$

Thus, the time (or phase) when squeezing occurs depends on the squeeze parameter, R.

Exercise 15.2 Derive Eq. (15.35).

The average electric field, Eq. (15.32), can be represented in the complex plane by a rotating phasor, as shown in Fig. 15.2a (with $r = 0$), and with the uncertainty represented by the gray oval. Figure 15.2 represents the case where $R < 0$. The average electric field and its uncertainty, shown in Fig. 15.2b, is obtained by projection of the phasor onto the imaginary axis. Note the negative sign in the amplitude of Eq. (15.32). Thus, $-\langle E(t) \rangle$ is given in Fig. 15.2b. As the phasor rotates, the uncertainty in electric field is smallest when $\omega t = 0, \pi, \ldots$ and greatest when $\omega t = \frac{\pi}{2}, \frac{3\pi}{2}, \ldots$, consistent with Eqs. (15.36) and (15.37), respectively. This case is called phase squeezing and corresponds to a reduction in the uncertainty of the P quadrature below the standard quantum limit (SQL) at certain times.

Conversely, Fig. 15.3 shows the case for $R > 0$. Here, the uncertainty is smallest when $\omega t = \frac{\pi}{2}, \frac{3\pi}{2}, \ldots$ and greatest when $\omega t = 0, \pi, \ldots$, corresponding to Eqs. (15.38)

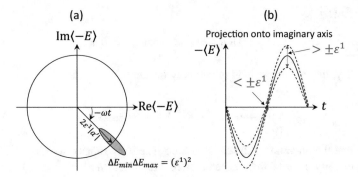

Fig. 15.2 (a) Phasor representation of squeezed field if $R < 0$. (b) Resulting average electric field, $-\langle E(t) \rangle$, and its uncertainty (dashed lines and red arrows)

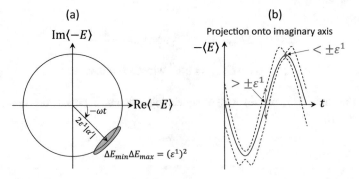

Fig. 15.3 (a) Phasor representation of squeezed field if $R > 0$. (b) Resulting average electric field, $-\langle E(t) \rangle$, and its uncertainty (dashed lines and red arrows)

and (15.39), respectively. This case is called amplitude squeezing or Q quadrature squeezing and corresponds to a reduction in the uncertainty of the Q quadrature below the SQL at certain times.

Thus, we can measure either the phase or the amplitude below the SQL, but only for a short period of time. As we will see in Chap. 16, we need homodyne detection (our "strobe light") to take advantage of the quadrature squeezing.

15.5   Quadratures

The Q quadrature of squeezed light is

$$Q = \langle Q \rangle = \langle \alpha, R | \widehat{Q} | \alpha, R \rangle \tag{15.40}$$

where

$$\widehat{Q} = \frac{1}{\sqrt{2}} \left(\widehat{a} + \widehat{a}^\dagger \right) \tag{15.41}$$

Using Eq. (15.18), we get

$$Q = \frac{1}{\sqrt{2}} \langle \alpha | \widehat{S}^\dagger \left(\widehat{a} + \widehat{a}^\dagger \right) \widehat{S} | \alpha \rangle \tag{15.42}$$

Using Eqs. (15.23) and (15.26), we get

$$Q = \frac{1}{\sqrt{2}} \langle \alpha | [(\widehat{a} \cosh R - \widehat{a}^\dagger \sinh R) + (\widehat{a}^\dagger \cosh R - \widehat{a} \sinh R)] | \alpha \rangle \tag{15.43}$$

Using the definition of a coherent state in Eq. (10.1) gives

$$Q = \frac{1}{\sqrt{2}} [(\alpha \cosh R - \alpha^* \sinh R) + (\alpha^* \cosh R - \alpha \sinh R)] \tag{15.44}$$

Using Eqs. (15.24) and (15.25), we get

$$Q = \frac{1}{\sqrt{2}} e^{-R} (\alpha + \alpha^*) \tag{15.45}$$

If $\alpha \in \mathbb{R}$,

$$Q = \frac{1}{\sqrt{2}} e^{-R}(2\alpha) \qquad (15.46)$$

It is straightforward to show that (Exercise 15.3)

$$\langle Q^2 \rangle = \langle \alpha, R | \widehat{Q}^2 | \alpha, R \rangle = \frac{1}{2} e^{-2R}(2\alpha)^2 + \frac{1}{2} e^{-2R} \qquad (15.47)$$

which gives the uncertainty

$$\Delta Q = \sqrt{\langle Q^2 \rangle - \langle Q \rangle^2} = \frac{1}{\sqrt{2}} e^{-R} \qquad (15.48)$$

Similarly, the uncertainty in the P quadrature is

$$\Delta P = \frac{1}{\sqrt{2}} e^{R} \qquad (15.49)$$

Recall that the quadratures for the coherent state were $\Delta Q = \Delta P = \frac{1}{\sqrt{2}}$, representing a minimum uncertainty state or standard quantum limit (SQL). For squeezed light with $R < 0$, we have an exponential increase (e^{-R}) in Q but an exponential decrease (e^{R}) in P compared to the SQL, and vice versa for $R > 0$. However, the product of the uncertainties is the same as that of the minimum uncertainty state; that is, the uncertainty relation is still satisfied:

$$\Delta Q \Delta P = \frac{1}{2} \qquad (15.50)$$

The quadrature representation of the state is shown in Fig. 15.4 for $R < 0$, and in Fig. 15.5 for $R > 0$. In the coherent state, the uncertainties are distributed equally between the two quadratures. In the squeezed state, the uncertainty in one quadrature is decreased at the expense of increasing the other. Recalling the number-phase

Fig. 15.4 Quadrature representation for phase squeezing

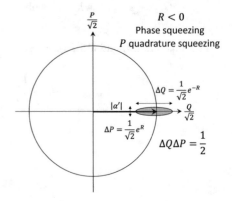

Fig. 15.5 Quadrature representation for amplitude squeezing

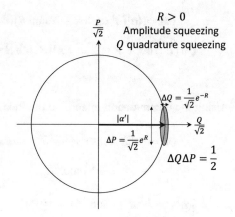

$R > 0$
Amplitude squeezing
Q quadrature squeezing

$$\Delta Q = \frac{1}{\sqrt{2}}e^{-R}$$

$$\Delta P = \frac{1}{\sqrt{2}}e^{R}$$

$$\Delta Q \Delta P = \frac{1}{2}$$

uncertainty relation, we see that the noise is redistributed between the amplitude and phase of the field.

Exercise 15.3 Derive Eq. (15.47).

15.6 Squeezed Power

The uncertainty of certain quadratures for squeezed light is reduced compared to that of coherent light. However, it is the signal-to-noise ratio (SNR) that really matters. Let us check the power in the beam for squeezed light versus coherent light, and then calculate the SNR. For a coherent state of amplitude α' we have

$$\langle n \rangle = \langle \alpha' | \widehat{N} | \alpha' \rangle = |\alpha'|^2 \tag{15.51}$$

For a squeezed state, we have

$$\langle n \rangle = \langle \alpha, R | \widehat{N} | \alpha, R \rangle \tag{15.52}$$

$$= \langle \alpha, R | \widehat{a}^\dagger \widehat{a} | \alpha, R \rangle \tag{15.53}$$

$$= \langle \alpha | \widehat{S}^\dagger \widehat{a}^\dagger \widehat{a} \widehat{S} | \alpha \rangle \tag{15.54}$$

Since $\widehat{S}$ is unitary ($\widehat{S}\widehat{S}^\dagger = 1$), we can write

$$\langle n \rangle = \langle \alpha | \widehat{S}^\dagger \widehat{a}^\dagger \widehat{S}\widehat{S}^\dagger \widehat{a} \widehat{S} | \alpha \rangle = \langle \alpha | \left(\widehat{S}^\dagger \widehat{a}^\dagger \widehat{S} \right) \left(\widehat{S}^\dagger \widehat{a} \widehat{S} \right) | \alpha \rangle \tag{15.55}$$

Using Eqs. (15.23) and (15.26)

$$\langle n \rangle = \langle \alpha | (\hat{a}^\dagger \cosh R - \hat{a} \sinh R)(\hat{a} \cosh R - \hat{a}^\dagger \sinh R) | \alpha \rangle \qquad (15.56)$$

$$= \langle \alpha | \left(\hat{a}^\dagger \hat{a} \cosh^2 R - \hat{a}^\dagger \hat{a}^\dagger \cosh R \sinh R - \hat{a}\hat{a} \sinh R \cosh R + \hat{a}\hat{a}^\dagger \sinh^2 R \right) | \alpha \rangle$$

$$(15.57)$$

Using the commutation relation to produce normal ordering gives

$$\langle n \rangle = \langle \alpha | [\hat{a}^\dagger \hat{a} \cosh^2 R - \hat{a}^\dagger \hat{a}^\dagger \cosh R \sinh R - \hat{a}\hat{a} \sinh R \cosh R \qquad (15.58)$$

$$+ \left(\hat{a}^\dagger \hat{a} + 1 \right) \sinh^2 R] | \alpha \rangle$$

$$= |\alpha|^2 \left(\cosh^2 R + \sinh^2 R \right) - \left[(\alpha^*)^2 + (\alpha)^2 \right] \cosh R \sinh R + \sinh^2 R \quad (15.59)$$

Using Eq. (15.31) and assuming $\alpha \in \mathbb{R}$, we get

$$\langle n \rangle = (\alpha')^2 + \sinh^2 R \qquad (15.60)$$

For typical values of R (~2 − 3), $\sinh^2 R \ll (\alpha')^2$, and we get

$$\langle n \rangle \sim (\alpha')^2 \qquad (15.61)$$

which is the same as Eq. (15.51). Thus, a coherent state and squeezed state of the same amplitude α' have approximately the same power in the beam. However, the uncertainty of the squeezed state is reduced compared to that of the coherent state. Thus, the SNR of the squeezed state is improved compared to a coherent state of the same power.

15.7 Fragility of Squeezing

The reason we do not use squeezing routinely is due to the fragility of the squeezed state. The quadrature squeezing is easily lost due to absorption, scattering, or other loss mechanisms. We can represent these losses by a beam splitter, as shown in Fig. 15.6. A fraction of the squeezed light is lost (represented by the reflection), while the remainder continues (represented by the transmission). The operator for the transmitted light is

$$\hat{a}_4 = t\hat{a}_1 - r\hat{a}_2 \qquad (15.62)$$

The Q quadrature at D_4 is

Fig. 15.6 Beam splitter
model of optical losses

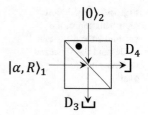

$$\langle Q_4 \rangle = {}_2\langle 0|_1\langle \alpha,\ R|\frac{1}{\sqrt{2}}\left(\hat{a}_4 + \hat{a}_4^\dagger\right)|\alpha,\ R\rangle_1|0\rangle_2 = \frac{1}{\sqrt{2}}e^{-R}(2\alpha)t \qquad (15.63)$$

which is the same as Eq. (15.46) but multiplied by the transmission coefficient, t. The
uncertainty can be calculated, giving:

$$\Delta Q_4 = \sqrt{\langle Q_4^2 \rangle - \langle Q_4 \rangle^2} = \frac{1}{\sqrt{2}}\left(t^2 e^{-2R} + r^2\right)^{1/2} \qquad (15.64)$$

Note that we recover Eq. (15.48) when we set $t = 1$ and $r = 0$ in Eq. (15.64) for a
lossless beam splitter. We see that the squeezing is reduced due to the r^2 term in
Eq. (15.64), representing the optical loss. The r^2 term arises due to the Q quadrature
of the vacuum fluctuations in port 2! For this reason, the squeezing is easily
destroyed unless great care is taken to remove optical losses.

Exercise 15.4 Derive Eqs. (15.63) and (15.64).

15.8   Squeezed Vacuum

As mentioned earlier, the optical parametric amplification process also works for the
vacuum state, resulting in a reduction (squeezing) of the uncertainty of the electric
field of the vacuum state. The squeezing operator applied to the vacuum produces the
"squeezed vacuum", that is, a coherent state with $\alpha = 0$:

$$\hat{S}(R)|0\rangle = |0,\ R\rangle \qquad (15.60)$$

Based on the preceding results, the properties of squeezed vacuum include

$$\langle E \rangle = \langle 0,\ R|\hat{E}(r,\ t)|0,\ R\rangle = 0 \qquad (15.61)$$

$$\langle Q \rangle = \langle 0,\ R|\hat{Q}|0,\ R\rangle = 0 \qquad (15.62)$$

$$\langle P \rangle = \langle 0,\ R|\hat{P}|0,\ R\rangle = 0 \qquad (15.63)$$

Fig. 15.7 Quadrature representation of the vacuum state $|0\rangle$ (dashed circle) and the squeezed vacuum state $|0, R\rangle$ (gray oval)

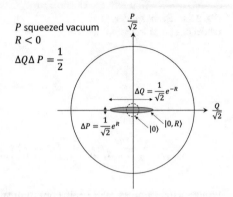

$$\Delta Q = \frac{1}{\sqrt{2}} e^{-R}, R < 0 \tag{15.64}$$

$$\Delta P = \frac{1}{\sqrt{2}} e^{R}, R < 0 \tag{15.65}$$

$$\langle n \rangle = \langle 0, R | \hat{a}^{\dagger} \hat{a} | 0, R \rangle = \sinh^{2} R \tag{15.66}$$

Notably, the average photon number of the squeezed vacuum is not zero. The quadrature representation of the P squeezed vacuum ($R < 0$) is shown in Fig. 15.7.

15.9   Photon Number Distribution of Squeezed Light

We saw before that the photon number distribution is Poissonian for the coherent state. Since the uncertainty of the phasor along the radius in Fig. 15.4 is larger than that for a coherent state, we have a super-Poissonian distribution in the photon number for the phase squeezed state. Conversely, the uncertainty in the phasor along the radius in Fig. 15.5 for the amplitude squeezed state is reduced compared to that for a coherent state, giving a sub-Poissonian photon number distribution.

Let us derive the photon number distribution of the squeezed vacuum state. We start with the vacuum state, which satisfies

$$\hat{a}|0\rangle = 0 \tag{15.67}$$

Applying the squeeze operator to Eq. (15.67), and using the fact that the squeeze operator is unitary ($\hat{S}^{\dagger} \hat{S} = 1$), gives

$$\widehat{S}\widehat{a}\widehat{S}^\dagger\widehat{S}|0\rangle = 0 \tag{15.68}$$

or, using Eq. (15.60)

$$\widehat{S}\widehat{a}\widehat{S}^\dagger|0, \ R\rangle = 0 \tag{15.69}$$

Similar to the derivation of Eq. (15.23), using the Baker–Hausdorff formula, we have

$$\widehat{S}\widehat{a}\widehat{S}^\dagger = \widehat{a}\cosh R + \widehat{a}^\dagger \sinh R \tag{15.70}$$

Thus, the squeezed vacuum satisfies the eigenvalue equation:

$$\left(\widehat{a}\cosh R + \widehat{a}^\dagger \sinh R\right)|0, \ R\rangle = 0 \tag{15.71}$$

To determine the photon number distribution, we want to express the vacuum state as a superposition of Fock states:

$$|0, \ R\rangle = \sum_{n=0}^{\infty} c_n|n\rangle \tag{15.72}$$

Then, the photon number distribution will be given by the probabilities, $|c_n|^2$. If we substitute Eq. (15.72) into the eigenvalue equation in Eq. (15.71), we get

$$\left(\widehat{a}\cosh R + \widehat{a}^\dagger \sinh R\right)\sum_{n=0}^{\infty} c_n|n\rangle = 0 \tag{15.73}$$

Equation (15.73) gives the recursion relation

$$c_{n+1} = -\tanh R\sqrt{\frac{n}{n+1}}\, c_{n-1} \tag{15.74}$$

The probability of the squeezed vacuum state having an odd number of photons is zero. This is not surprising because photons are created or annihilated in pairs from the vacuum according to the optical parametric process. We start from the vacuum and the squeezing operator adds or subtracts two photons at a time. Only the even numbered coefficients contain the vacuum state. The solution to Eq. (15.74) for even solutions is

$$c_{2n} = (-1)^n (\tanh R)^n \left[\frac{(2n-1)!!}{(2n)!!} \right]^{1/2} c_0 \qquad (15.75)$$

where n is an integer, and !! denotes the double factorial. The double factorial, for example $n!!$, is the product of all the integers from 1 up to n that have the same parity (odd or even) as n. For example, if n is even, then $n!! = n(n-2)(n-4)\ldots(4)(2)$.
c_0 can now be determined from the normalization condition

$$\sum_{n=0}^{\infty} |c_{2n}|^2 = 1 \qquad (15.76)$$

resulting in

$$|c_0|^2 \left(1 + \sum_{n=1}^{\infty} \frac{(\tanh R)^{2n} (2n-1)!!}{(2n)!!} \right) = 1 \qquad (15.77)$$

Next, we use the following (not obvious!) identities:

$$1 + \sum_{n=1}^{\infty} \frac{(z)^n (2n-1)!!}{(2n)!!} = (1-z)^{-1/2} \qquad (15.78)$$

$$(2n)!! = 2^n n! \qquad (15.79)$$

$$(2n-1)!! = \frac{2^{-n}(2n)!}{n!} \qquad (15.80)$$

which gives

$$c_0 = \frac{1}{\sqrt{\cosh R}} \qquad (15.81)$$

and

$$c_{2n} = (-1)^n \frac{\sqrt{(2n)!}}{2^n n!} \frac{(\tanh R)^n}{\sqrt{\cosh R}} \qquad (15.82)$$

From Eq. (15.72), we get

$$|0, R\rangle = \sum_{n=0}^{\infty} (-1)^n \frac{\sqrt{(2n)!}}{2^n n!} \frac{(\tanh R)^n}{\sqrt{\cosh R}} |2n\rangle \qquad (15.83)$$

Thus, the probability of detecting $2n$ photons from the vacuum squeezed state is given by the square of the probability amplitude in Eq. (15.83):

Fig. 15.8 Photon number distribution for squeezed vacuum state from Eq. (15.84) with $R = 2$

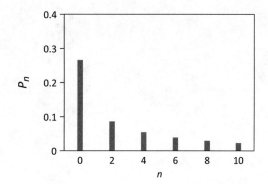

$$P_{2n} = \frac{(2n)!}{2^{2n}(n!)^2} \frac{(\tanh R)^{2n}}{\cosh R} \tag{15.84}$$

Figure 15.8 shows an example of the photon number distribution from Eq. (15.84) for a squeezed vacuum with $R = 2$.

The squeezed coherent state can be obtained by applying the squeeze operator to the vacuum, $\widehat{S}(R)|0\rangle = |0, R\rangle$, and then applying the displacement operator introduced in Sect. 10.9, $\widehat{D}(\alpha)\widehat{S}(R)|0\rangle = |\alpha, R\rangle$. This results in a recursion relation that can be solved for the photon number distribution, similar to the previous procedure for the squeezed vacuum. The squeezed coherent state contains a mixture of even and odd photon numbers with the possibility of sub-Poissonian statistics. A derivation can be found in Refs. [2–4].

References

1. R.E. Slusher et al., *Observation of squeezed states generated by four-wave mixing in an optical cavity*, Phys. Rev. Lett. 55 (1985) 2409.
2. C.C. Gerry and P.L. Knight, *Introductory quantum optics* (2005, Cambridge University Press), pp. 160-165.
3. R.W. Henry and S.C. Glotzer, *A squeezed state primer*, Amer. J. Phys. 56 (1988) 318.
4. J.J. Gong and P.K. Aravind, *Expansion coefficients of a squeezed coherent state in the number state basis*, Amer. J. Phys. 58 (1990) 1003.

Chapter 16
Squeezed Light in an Interferometer

The uncertainty and signal-to-noise ratio (SNR) of a homodyne signal using squeezed light in an interferometer is derived, showing SNR below the shot noise limit. The use of squeezed light for improved detection of gravitational waves is presented as a key application.

16.1 Homodyne Signal Using Squeezed Light

In Chap. 14, we learned that the SNR for the coherent state arises from the uncertainty in the field quadrature of the vacuum input to the interferometer. To improve the SNR, we can replace the vacuum, $|0\rangle$, in the interferometer with squeezed vacuum, $|0, R\rangle$! Using squeezed light, we will see that we can beat the shot noise limit in interferometry. This approach was proposed by Carlton M. Caves [1].

The input state to the interferometer in Fig. 16.1 is $|\psi_{\text{in}}\rangle = |0, R\rangle_1 |\alpha\rangle_2$. The local oscillator is the coherent state $|\alpha\rangle_2$ on input port 2, while $|0, R\rangle_1$ is the vacuum squeezed state on input port 1. The operator for the homodyne signal from Chap. 14 is

$$\hat{n}_4 - \hat{n}_3 = -\sin\varepsilon\left(\hat{a}_1^\dagger\hat{a}_1 - \hat{a}_2^\dagger\hat{a}_2\right) - i\cos\varepsilon\left(\hat{a}_1^\dagger\hat{a}_2 - \hat{a}_2^\dagger\hat{a}_1\right) \qquad (16.1)$$

Retaining the non-zero terms gives

$$\langle n_4\rangle - \langle n_3\rangle = \sin\varepsilon\,_2\langle\alpha|\,_1\langle 0,\ R|(|\alpha|^2 - \hat{a}_1^\dagger\hat{a}_1)|0,\ R\rangle_1|\alpha\rangle_2 \qquad (16.2)$$

From Eq. (15.66), we get

© The Author(s), under exclusive license to Springer Nature Switzerland AG 2022
R. LaPierre, *Getting Started in Quantum Optics*, Undergraduate Texts in Physics,
https://doi.org/10.1007/978-3-031-12432-7_16

Fig. 16.1 Interferometer
with squeezed vacuum on
port 1 and coherent state on
port 2

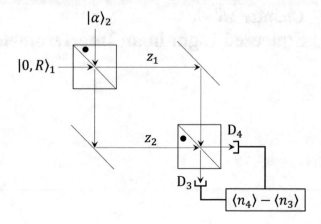

$$\langle n_4 \rangle - \langle n_3 \rangle = \sin \varepsilon \left(|\alpha|^2 - \sinh^2 R \right) \qquad (16.3)$$

For a strong coherent state, we have $|\alpha|^2 \gg \sinh^2 R$, which gives

$$\langle n_4 \rangle - \langle n_3 \rangle \sim |\alpha|^2 \sin \varepsilon \qquad (16.4)$$

$$= \langle n_2 \rangle \sin \varepsilon \qquad (16.5)$$

For a small phase change (small ε), we get

$$\langle n_4 \rangle - \langle n_3 \rangle \sim \langle n_2 \rangle \varepsilon \qquad (16.6)$$

which is the same result as that obtained in Chap. 14 for the coherent state.

Next, we evaluate $\langle (n_4 - n_3)^2 \rangle$. Assuming small displacements in the interferometer ($\varepsilon \to 0$) and α real, we get

$$\langle (n_4 - n_3)^2 \rangle = -\alpha^2 \,_1\langle 0, \ R | (\hat{a}_1^\dagger - \hat{a}_1)^2 |0, \ R \rangle_1 \qquad (16.7)$$

$$= \alpha^2 e^{2R} \qquad (16.8)$$

$$= e^{2R} \langle n_2 \rangle \qquad (16.9)$$

Exercise 16.1 Derive Eqs. (16.7) and (16.8).

From Eqs. (16.6) and (16.9), the uncertainty is

$$\Delta(n_4 - n_3) = \sqrt{\left\langle (n_4 - n_3)^2 \right\rangle} - \langle n_4 - n_3 \rangle \sim e^R \sqrt{\langle n_2 \rangle} \qquad (16.10)$$

The shot noise is reduced by a factor e^R for $R < 0$. The SNR is

$$\text{SNR} = \frac{\varepsilon \langle n_2 \rangle}{e^R \sqrt{\langle n_2 \rangle}}$$

$$= \varepsilon \sqrt{\langle n_2 \rangle}\, e^{-R} \qquad (16.11)$$

Compared to Eq. (14.48), the SNR is improved by a factor e^{-R} for $R < 0$ due to the P quadrature squeezing (phase squeezing) of the light on port 1.

Exercise 16.2 Derive Eq. (16.10).

Today, noise suppression in excess of 10 dB has been achieved, corresponding to a factor of 10 reduction in noise, or $R{\sim}2.3$. Most squeezing experiments are based on optical parametric amplification (OPA). Figure 16.2 illustrates the homodyne signal measured for various states of light in one of the earlier squeezing experiments [2].

16.2   Laser Interferometer Gravitational-Wave Observatory (LIGO)

The most spectacular application of squeezed light is the Laser Interferometer Gravitational-wave Observatory (LIGO), located in Hanford, Washington, (Fig. 16.3) and Livingston, Louisiana, in the USA [3]. These two facilities each use a Michelson interferometer to measure displacements of space itself associated with passing gravitational waves. On Sept 14, 2015, the LIGO detectors saw space "vibrate" due to gravitational waves generated by the merger of two black holes as predicted by Einstein's general theory of relativity [4]. The 2017 Nobel Prize in Physics was awarded to Rainer Weiss, Barry Barish, and Kip Thorne for their work on gravitational wave detection.

LIGO is a Michelson interferometer with 4 km long arms. Gravitational waves cause a displacement of the interferometer mirrors, resulting in a change in phase ε:

$$\varepsilon = k\Delta x \qquad (16.12)$$

or, after rearranging,

$$\Delta x = \frac{\lambda \varepsilon}{2\pi} \qquad (16.13)$$

Fig. 16.2 Experimentally
measured electric field
versus time (homodyne
signal) for (from the top) the
coherent state, phase-
squeezed state, squeezing at
a phase of $\varphi = 48°$,
amplitude-squeezed state
and squeezed vacuum state.
(Reprinted by permission
from Springer Nature,
G. Breitenbach, S. Schiller
and J. Mlynek, Nature 387
(1997) 471 [2])

Fig. 16.3 Aerial photograph of the LIGO Hanford facility, showing the 4 km long arms of the interferometer. (Credit: Wikimedia Commons [3])

Thus, the minimum displacement Δx that can be measured is determined by the minimum uncertainty in ε, assuming the wavelength is well known.

We can determine a phase change, ε, from a change in some measurement parameter, M, and the sensitivity of the measurement parameter to the phase, $\frac{dM}{d\varepsilon}$. Thus,

$$\varepsilon = \frac{M}{\frac{dM}{d\varepsilon}} \tag{16.14}$$

The minimum phase we can measure corresponds to the minimum possible value of M that can be measured:

$$\varepsilon_{min} = \frac{M_{min}}{\frac{dM}{d\varepsilon}} \tag{16.15}$$

The minimum value of M that can be measured corresponds to its uncertainty, ΔM:

$$\varepsilon_{min} = \frac{\Delta M}{\frac{dM}{d\varepsilon}} \tag{16.16}$$

Let us suppose that M corresponds to a homodyne measurement. For a coherent state in an interferometer, we found the homodyne signal was given by Eq. (14.32):

$$M = \langle n_4 \rangle - \langle n_3 \rangle = \langle n_2 \rangle \varepsilon \tag{16.17}$$

The sensitivity is therefore

$$\frac{dM}{d\varepsilon} = \langle n_2 \rangle \tag{16.18}$$

The uncertainty was given by the shot noise limit (Eq. (14.47)):

$$\Delta M = \Delta(n_4 - n_3) \sim \sqrt{\langle n_2 \rangle} \tag{16.19}$$

Finally, according to Eq. (16.13):

$$\varepsilon_{min} = \frac{\Delta M}{\frac{dM}{d\varepsilon}} = \frac{\sqrt{\langle n_2 \rangle}}{\langle n_2 \rangle} = 1/\sqrt{\langle n_2 \rangle} \tag{16.20}$$

Note that Eq. (16.20) can also be derived from the number-phase uncertainty relation, $\Delta n \Delta \varepsilon = 1$, which gives $\Delta \varepsilon = \frac{1}{\Delta n} = \frac{1}{\sqrt{\langle n_2 \rangle}}$ for the shot noise limit.

LIGO uses a wavelength of $\lambda = 1$ μm and optical power of 100 kW in the interferometer arms, corresponding to a photon flux of $\sim 10^{24}$ photons/s. Assuming a measurement time of 1 s, this corresponds to $\langle n_2 \rangle \sim 10^{24}$. Hence, using Eq. (16.13), the minimum displacement that can be measured is

$$\Delta x_{\min} = \frac{\lambda}{2\pi\sqrt{\langle n_2 \rangle}} = \frac{(10^{-6}\ \text{m})}{2\pi\sqrt{10^{24}}} \sim 10^{-19}\ \text{m} \qquad (16.21)$$

or about 10^4 times smaller than the diameter of a proton! For the 4 km path length in the LIGO interferometer, this corresponds to a relative displacement (strain) of

$$\frac{\Delta x}{x} \sim \frac{10^{-19}\ \text{m}}{4000\ \text{m}} \sim 10^{-22} \qquad (16.22)$$

Implementing squeezed light on the empty port of the interferometer can reduce the uncertainty by a factor of e^R ($R < 0$) according to Eq. (16.10) and improve the strain sensitivity in Eq. (16.22) by a factor e^{-R}. The resulting improvement in strain sensitivity is shown in Fig. 16.4 [5, 6]. The use of squeezed light in LIGO has significantly increased the rate of observed gravitational-wave events.

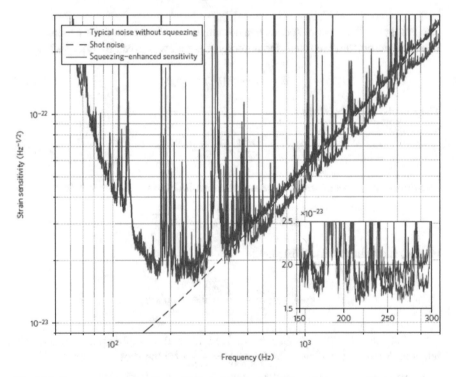

Fig. 16.4 Improvements in strain sensitivity in LIGO achieved by using squeezed light. The sharp lines are due to parasitic effects such as mechanical resonances. (Reprinted by permission from Springer Nature, J. Aasi et al., Nature Photonics 7 (2013) 613 [6])

> **Exercise 16.3** Investigate and discuss other potential applications of squeezed light.

References

1. C.M. Caves, *Quantum-mechanical noise in an interferometer*, Phys. Rev. D 23 (1981) 1693.
2. G. Breitenbach, S. Schiller and J. Mlynek, *Measurement of the quantum states of squeezed light*, Nature 387 (1997) 471.
3. https://web.archive.org/web/20170614214509/http://www.ligo.org/multimedia/gallery/lho-images/Aerial5.jpg
4. B.P. Abbott et al., *Observation of gravitational waves from a binary black hole merger*, Phys. Rev. Lett. 116 (2016) 061102.
5. J. Gea-Banacloche and G. Leuchs, *Squeezed states for interferometric gravitational-wave detectors*, J. Mod. Opt. 34 (1987) 793.
6. J. Aasi et al., *Enhanced sensitivity of the LIGO gravitational wave detector by using squeezed states of light*, Nature Photonics 7 (2013) 613.

Chapter 17
Heisenberg Limit

The uncertainty in photon number of a coherent source leads to the shot noise limit. We could eliminate this uncertainty by using a Fock state, which has a definite photon number. This approach produces the "Heisenberg limit", which is the ultimate limit on measurement uncertainty. Multiphoton entangled states (called NOON states) in an interferometer are introduced as a means of achieving the Heisenberg limit. Super-sensitivity and super-resolution are introduced as applications. Methods of producing NOON states are described, and we introduce the Hong–Ou–Mandel (HOM) effect and linear optical quantum state engineering.

17.1 Heisenberg Limit

As discussed in Chap. 16, we found the minimum phase that can be measured is given by

$$\varepsilon_{\min} = \frac{\Delta M}{\frac{dM}{d\varepsilon}} \qquad (17.1)$$

where ΔM is the uncertainty in a measurement parameter, M, and $\frac{dM}{d\varepsilon}$ is the sensitivity of ε with respect to M. For homodyne detection of a coherent state in an interferometer, we found in the previous chapter that $\Delta M = \sqrt{\langle n \rangle}$ and $\frac{dM}{d\varepsilon} = \langle n \rangle$, giving $\varepsilon_{\min} = \frac{1}{\sqrt{\langle n \rangle}}$, which is the shot noise limit. This can be improved to $\varepsilon_{\min} = e^R / \sqrt{\langle n \rangle}$ by using squeezed light (with $R < 0$). Can we do any better than this?

Let us suppose that the smallest measurement we can make is $\Delta M = 1$, corresponding to the measurement of a single photon. The minimum possible phase measurement is then

© The Author(s), under exclusive license to Springer Nature Switzerland AG 2022
R. LaPierre, *Getting Started in Quantum Optics*, Undergraduate Texts in Physics,
https://doi.org/10.1007/978-3-031-12432-7_17

$$\varepsilon_{\min} = \frac{\Delta M}{\frac{dM}{d\varepsilon}} = \frac{1}{\langle n \rangle} \tag{17.4}$$

Equation (17.4) is called the Heisenberg limit, which is much better than the $1/\sqrt{\langle n \rangle}$ value of the shot noise limit! One can also obtain Eq. (17.4) from the number-phase uncertainty relation, $\Delta \varepsilon \Delta n = 1$. If we have $\langle n \rangle$ photons, the maximum uncertainty is $\Delta n = \langle n \rangle$. Then, from the number-phase uncertainty relation, we get $\varepsilon_{\min} = \Delta \varepsilon = \frac{1}{\Delta n} = \frac{1}{\langle n \rangle}$, which is the same as Eq. (17.4).

For example, as discussed in the previous chapter, the LIGO interferometer with $\langle n \rangle = 10^{24}$ photons and shot-noise limited phase uncertainty gave $\varepsilon_{\min} = \frac{1}{\sqrt{\langle n \rangle}} = 10^{-12}$, corresponding to a minimum displacement measurement of $\Delta x \sim 10^{-19}$ m. In the Heisenberg limit, the phase uncertainty would be $\varepsilon_{\min} = \frac{1}{\langle n \rangle} = 10^{-24}$, corresponding to a minimum displacement measurement of $\Delta x \sim 10^{-31}$ m, a 12 order of magnitude improvement! This scale is approaching the Planck scale of 10^{-35} m where quantum gravity and the graininess of space-time become relevant! In the following sections, we examine approaches to achieving the Heisenberg limit.

17.2 Phase Shifter

A phase shifter is an optical device that imparts a phase shift, $e^{i\varphi}$, to a classical field. A phase shifter could simply be a piece of glass of certain thickness and refractive index, or it could correspond to an optical path length difference (e.g., in an interferometer). As shown in Fig. 17.1a, a phase shifter will impart a phase shift, $e^{i\varphi}$, to a coherent state, $|\alpha\rangle$, which resembles a classical field. However, something very different happens for the number state. As shown in Fig. 17.1b, a number state undergoes a phase shift of $e^{in\varphi}$ in passing through a phase shifter. Let us explore how we might use this effect.

Exercise 17.1 Investigate and explain why a number state undergoes a phase shift of $e^{in\varphi}$ in passing through a phase shifter, as compared to $e^{i\varphi}$ for a coherent state.

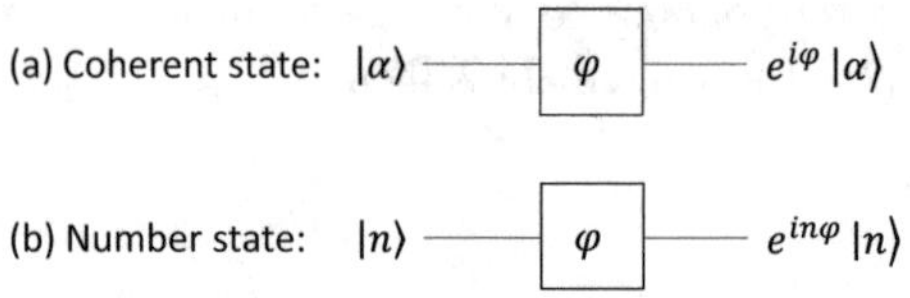

Fig. 17.1 The effect of a phase shifter on (**a**) a coherent state and (**b**) a number state

(a) Coherent state: $|\alpha\rangle$ —— $\boxed{\varphi}$ —— $e^{i\varphi}|\alpha\rangle$

(b) Number state: $|n\rangle$ —— $\boxed{\varphi}$ —— $e^{in\varphi}|n\rangle$

17.3 NOON States

Consider an interferometer where one path of the interferometer contains a phase shifter. For this analysis, we adopt a simplified view of the interferometer, as shown in Fig. 17.2, where the input and output optics (mirrors and beam splitters) are consolidated into boxes on the left and right. We suppose we can create a state in the two paths of the interferometer where we have n photons that are either all in the upper mode (input path 1 before the phase shifter) and none in the lower mode 2, or vice versa (i.e., we have a superposition of the two possibilities):

$$|\psi\rangle = \frac{|n, 0\rangle + |0, n\rangle}{\sqrt{2}} \tag{17.5}$$

The first number in each of the Dirac brackets in Eq. (17.5) denotes the number of photons in the upper path, and the second number denotes the number of photons in the lower path. This state is a superposition state of $|n, 0\rangle$ and $|0, n\rangle$. This is called a "NOON" state when $n \geq 2$. Note that the NOON state is an entangled state; that is, it is a nonseparable state. As n becomes large, the NOON state is called a "Schrodinger cat state". The NOON state was first discussed in 1989 by Barry Sanders in the context of decoherence of Schrodinger cat states [1].

In the upper path of the interferometer, we insert a phase shifter. According to Fig. 17.1b, the resulting state after the phase shifter is

$$|\psi\rangle = \frac{e^{in\varphi}|n, 0\rangle + |0, n\rangle}{\sqrt{2}} \tag{17.6}$$

where $\varphi = k\Delta x$ represents an optical path length difference between the two modes.

17.4 Super-sensitivity

Suppose we measure a homodyne signal, $\langle n_4 - n_3 \rangle$, at the output of the interferometer in Fig. 17.2. For a coherent state in an interferometer, we previously found (Eq. (14.26))

Fig. 17.2 Effect of phase shifter on a NOON state in an interferometer

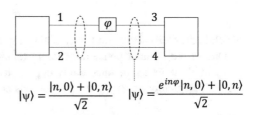

Fig. 17.3 Homodyne signal for a coherent state (red) versus a NOON state with $n = 3$ (green) in an interferometer

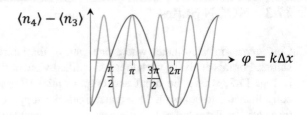

$$\langle n_4 \rangle - \langle n_3 \rangle = -\langle n \rangle \cos(\varphi) \tag{17.7}$$

where $\langle n \rangle$ is the average number of photons in the coherent state. Equation (17.7) is represented by the red curve in Fig. 17.3. If we operate near the quadrature with $\varphi = \frac{\pi}{2} + \varepsilon$, we previously obtained, for small ε,

$$\langle n_4 \rangle - \langle n_3 \rangle = \langle n \rangle \sin(\varepsilon) \sim \langle n \rangle \, \varepsilon \tag{17.8}$$

For a NOON state, Eq. (17.7) changes to

$$\langle n_4 \rangle - \langle n_3 \rangle = -n \cos(n\varphi) \tag{17.9}$$

where n is the number of photons in the NOON state. Figure 17.3 illustrates Eq. (17.9) for $n = 3$. If we operate near $\varphi = \frac{\pi}{2} + \varepsilon$, we obtain for small ε:

$$\langle n_4 \rangle - \langle n_3 \rangle = n \sin(n\varepsilon) \sim n \, (n\varepsilon) \tag{17.10}$$

We see that the slope at the horizontal crossings (e.g., $\frac{\pi}{2}$) for the number state increases by a factor of n compared to the coherent state. Thus, comparing Eq. (17.8) and (17.10), we have an improvement in phase sensitivity by a factor n. This improvement is called "super-sensitivity".

With the NOON state, we need to detect $\langle n_4 \rangle - \langle n_3 \rangle = n$ photons at a time:

$$\langle n_4 \rangle - \langle n_3 \rangle = n \tag{17.11}$$

or, from Eq. (17.10):

$$\varepsilon = \frac{1}{n} \tag{17.12}$$

which is the Heisenberg limit, identical to Eq. (17.4).

Due to the number-phase uncertainty relation, we expect the minimum uncertainty in phase when the uncertainty in photon number is greatest. In NOON states, we are completely uncertain if all n photons are in mode 3 (and none in 4) or if all

n photons are in mode 4 (and none in 3). This gives the maximum uncertainty in the number of photons of $\Delta n = n$. Thus, from the number-phase uncertainty relation,

$$\Delta n \Delta \varepsilon = 1 \rightarrow \Delta \varepsilon = \frac{1}{\Delta n} = \frac{1}{n} \tag{17.13}$$

which is again the Heisenberg limit.

17.5   Super-resolution and Quantum Lithography

Figure 17.3 shows a reduction in wavelength from λ for a coherent state to λ/n for a NOON state in an interferometer. λ/n is referred to as the de Broglie wavelength of the photon. The de Broglie wavelength of a single photon is

$$\lambda = \frac{h}{p} = \frac{hc}{E} \tag{17.14}$$

The de Broglie wavelength of a biphoton is

$$\lambda = \frac{h}{p} = \frac{hc}{2E} \tag{17.15}$$

Continuing in this manner, the de Broglie wavelength of the NOON state is

$$\lambda = \frac{h}{p} = \frac{hc}{nE} \tag{17.16}$$

which is called a "superphoton" [2].

A superphoton with reduced de Broglie wavelength would allow improvements in the spatial resolution of imaging systems. In a conventional imaging system, the Rayleigh criterion states that diffraction limits the smallest separation between two object points that can be resolved in an optical image. Classical physics tells us that the minimum angular separation of two object points is $1.22\lambda/D$, where D is the diameter of the collecting aperture. "Super-resolution" achieves a spatial resolution in an optical imaging system using superphotons with reduced de Broglie wavelength that can beat the Rayleigh criterion. The main idea of sub-Rayleigh imaging is to replace intensity measurements with spatially resolved n photon detection using photon-number-resolving detectors.

However, NOON states suffer a significant weakness. The probability of detecting n photons arriving at the same place decreases exponentially with n due to scattering; that is, NOON states are exponentially sensitive to absorption. Beer's law states that classical or coherent light will decay exponentially ($e^{-\gamma}$) according to an absorption coefficient, γ. However, the NOON state absorption becomes $e^{-n\gamma}$. This reduces the slope at the horizontal crossings, killing the super-sensitivity advantage. Thus, super-resolution has only been demonstrated with a few photons [3].

One of the proposed applications of super-resolution is quantum lithography, where the superphoton with reduced de Broglie wavelength is used to expose photoresists for improved spatial resolution in lithography [4]. Besides the difficulty in creating NOON states, the problem with quantum lithography is that it has proven very difficult to make efficient *n*-photon-sensitive absorbing resists.

Other applications of quantum metrology include the entanglement of ion states, which could improve the accuracy of atomic clocks. We could measure a frequency to an accuracy of $\Delta\omega = \frac{\Delta\varphi}{t} = \frac{1}{tn}$, instead of $\Delta\omega = \frac{1}{t\sqrt{n}}$. We could also improve transit time measurements, distance measurements, and clock synchronization (used in GPS) [5].

17.6 Producing a NOON State

A superposition of one photon in mode 3 and none in mode 4 ($|1\rangle_3|0\rangle_4$), and one photon in mode 4 and none in mode 3 ($|0\rangle_3|1\rangle_4$), can be achieved using a beam splitter (Fig. 17.4). As we saw previously in Chap. 6, for a 50:50 beam splitter and single photon input on one of the ports, we get

$$|\psi_{\text{out}}\rangle = \frac{1}{\sqrt{2}}\left(|1\rangle_3|0\rangle_4 + |0\rangle_3|1\rangle_4\right) \tag{17.17}$$

This is an entangled state of a photon in mode 3 and mode 4. Technically, this is not a NOON state, since we need $n \geq 2$. The following sections will describe how to create a NOON state with $n \geq 2$.

Fig. 17.4 Producing a
NOON state with $n = 1$

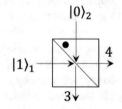

17.7   Hong–Ou–Mandel Effect

A NOON state with $n = 2$ can be produced by a single photon input on both port 1 and 2 of the beam splitter in Fig. 17.4:

$$|\psi_{\text{in}}\rangle = |1\rangle_1|1\rangle_2 \tag{17.18}$$

$$= \hat{a}_1^\dagger \hat{a}_2^\dagger |0\rangle_1 |0\rangle_2 \tag{17.19}$$

The output state is

$$|\psi_{\text{out}}\rangle = \left(r\hat{a}_3^\dagger + t\hat{a}_4^\dagger \right) \left(t\hat{a}_3^\dagger - r\hat{a}_4^\dagger \right) |0\rangle_3 |0\rangle_4 \tag{17.20}$$

$$= \left(rt\hat{a}_3^\dagger \hat{a}_3^\dagger - r^2 \hat{a}_3^\dagger \hat{a}_4^\dagger + t^2 \hat{a}_4^\dagger \hat{a}_3^\dagger - tr\hat{a}_4^\dagger \hat{a}_4^\dagger \right) |0\rangle_3 |0\rangle_4 \tag{17.21}$$

Recalling Eq. (2.127), we get

$$|\psi_{\text{out}}\rangle = \sqrt{2}rt|2\rangle_3|0\rangle_4 - r^2|1\rangle_3|1\rangle_4 + t^2|1\rangle_3|1\rangle_4 - \sqrt{2}tr|0\rangle_3|2\rangle_4 \tag{17.22}$$

For a 50:50 beam splitter ($r = t = \frac{1}{\sqrt{2}}$), Eq. (17.22) becomes

$$|\psi_{\text{out}}\rangle = \frac{1}{\sqrt{2}} \left(|2\rangle_3|0\rangle_4 - |0\rangle_3|2\rangle_4 \right) \tag{17.23}$$

Equation (17.23) represents a NOON state with $n = 2$. This arises due to destructive interference of the probability amplitudes for a single photon in both modes 3 and 4, destroying the $|1\rangle_3|1\rangle_4$ state. This interference is a quantum mechanical effect called the Hong–Ou–Mandel (HOM) effect, named after C.K. Hong, Z.Y. Ou, and L. Mandel [6]. The HOM effect results in only two-photon detection occurring randomly at either D_3 or D_4 as depicted in Fig. 17.5.

Perfect destructive interference of the $|1\rangle_3|1\rangle_4$ mode occurs only when the two photon inputs are indistinguishable in every respect, that is, when they have the same

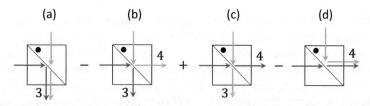

Fig. 17.5 The four possible photon paths leading to the Hong–Ou–Mandel (HOM) effect. The negative signs arise due to one of the reflections, leading to destructive interference of the amplitudes in (**b**) and (**c**). Only the outputs in (**a**) and (**d**) remain. The photon paths are labelled with different colors, although the photons are indistinguishable

Table 17.1 Comparison of quantum versus classical expectation for HOM experiment with 50:50 beam splitter

Mode 3	Mode 4	Quantum probability	Classical probability
$|2\rangle$	$|0\rangle$	$2|r|^2|t|^2 = 1/2$	$|r|^2|t|^2 = 1/4$
$|0\rangle$	$|2\rangle$	$2|r|^2|t|^2 = 1/2$	$|r|^2|t|^2 = 1/4$
$|1\rangle$	$|1\rangle$	$(|t|^2 - |r|^2)^2 = 0$	$|r|^4 + |t|^4 = 1/2$

frequency, same polarization, same coefficients (probability amplitudes) of the multimode state, and when they arrive at the same time on the beam splitter. Hence, the HOM effect can be used to measure the indistinguishability of two photons.

The classical expectation versus the quantum results for the HOM experiment are summarized in Table 17.1. Classically, we expect each of the fields to reflect from the beam splitter with probability $|r|^2$ and transmit with probability $|t|^2$. Conversely, the quantum probabilities are obtained from the square of the probability amplitudes in Eq. (17.21). The quantum result is very different than the classical one. Note that the probabilities sum to unity in either case. Classically, we expect coincidence counts with 50% probability. Therefore, the HOM effect should give coincidence counts with probability less than 50% to demonstrate quantum interference.

The experimental setup for the original HOM measurement is shown in Fig. 17.6a. The two photons were generated by parametric down-conversion in a KDP crystal. The two photons were redirected by mirrors (M1, M2) onto a beam splitter (BS) and two detectors (D1, D2) for coincidence counting. A time delay between the measurement at D1 and D2 is implemented by translation of the beam splitter. The results shown in Fig. 17.6b reveal nearly zero coincidence counts (the "HOM dip") for simultaneous measurement (corresponding to BS position near 300 μm) where the two photon modes overlap temporally. The temporal overlap of the two photons decreases by displacing the beam splitter. The coincidence counts increase as the wavepacket of the two photons no longer overlaps perfectly. A review of the HOM effect is available in Ref. [7].

17.8 High NOON State

A NOON state with large n is called a "high NOON state". Figure 17.7 illustrates a possible method for producing a high NOON state. The method, proposed in 2001 [8], uses two interferometers coupled by a Kerr material. The Kerr material uses the nonlinear Kerr effect to produce a π phase shift when one photon is incident from the upper interferometer but zero phase shift otherwise. The lower interferometer is set up so that all n photons input on mode 1 will go to the output of mode 3 and none to mode 4 when there is no phase shift from the Kerr material. However, when the Kerr material gives a π phase shift, all n photons in the lower interferometer will go in mode 4 and none in mode 3. The upper interferometer produces an entangled state with a superposition of 0 and 1 photon incident on the Kerr phase shifter, producing

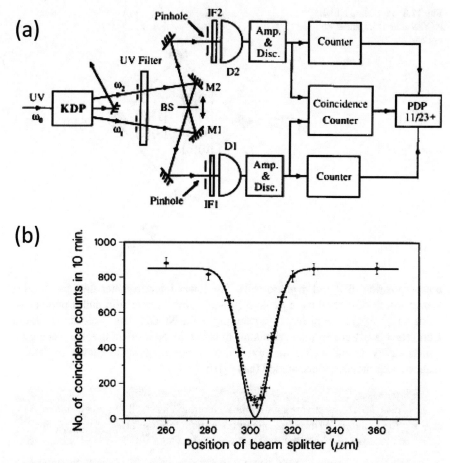

Fig. 17.6 (a) Experimental setup and (b) results for the original Hong–Ou–Mandel (HOM) measurement [6]. (Reprinted with permission from C.K. Hong, Z.Y. Ou and L. Mandel, Phys. Rev. Lett. 59 (1987) 2044, https://doi.org/10.1103/PhysRevLett.59.2044 [6]. Copyright 1987 by the American Physical Society)

Fig. 17.7 Producing a high NOON state using nonlinear optics (Kerr effect)

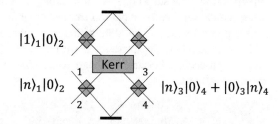

Fig. 17.8 Producing a high
NOON state from linear
optics

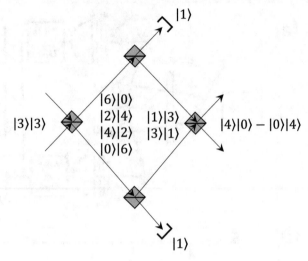

a superposition of 0 and π phase shift. The lower interferometer then produces a superposition of n photons in mode 3 ($|n\rangle_3|0\rangle_4$) for 0 phase shift and n photons in mode 4 ($|0\rangle_3|n\rangle_4$) for π phase shift, resulting in the NOON state. Unfortunately, the Kerr effect is extremely weak. Attempts to boost the Kerr effect involve placing an optical cavity around a Kerr material [9], or using an approach known as electromagnetically induced transparency (EIT) [10].

Exercise 17.2 Investigate and explain electromagnetically induced transparency (EIT) and describe its applications in quantum optics.

Another approach to producing high NOON states is to use entirely linear optics, as shown in Fig. 17.8. Due to a generalization of the HOM effect, an input state of $|3\rangle|3\rangle$ results in an output from the first beam splitter with a superposition of $|6\rangle|0\rangle$, $|2\rangle|4\rangle$, $|4\rangle|2\rangle$ and $|0\rangle|6\rangle$. If we perform a coincidence measurement on this state and obtain $|1\rangle|1\rangle$, then the measurement removes the $|6\rangle|0\rangle$ and $|0\rangle|6\rangle$ states from the superposition. The state collapses to a superposition of $|1\rangle|3\rangle$ and $|3\rangle|1\rangle$, since the measurement removed one photon from each mode. Passage through the final beam splitter produces the NOON state with $n = 4$, again due to the HOM effect (Exercise 17.3). This is a heralded process. If we obtain the $|1\rangle|1\rangle$ state upon measurement, then we proceed; otherwise, we start over. Thus, this approach will succeed with some probability.

Exercise 17.3 Show that the last beam splitter in Fig. 17.6 produces a NOON state with $n = 4$ from $|1\rangle|3\rangle + |3\rangle|1\rangle$.

17.9 Quantum State Engineering

The previous sections highlight the use of mirrors, phase shifters, and beam splitters to produce a desired quantum state. These devices are the resources available to realize a desired unitary transformation, together with photon counting measurements. Putting these elements together efficiently to realize a desired quantum state, which we call "quantum state engineering", is an active research area for all-optical quantum information processing. Knill et al. [11] have shown how linear optics together with photon counting measurements can enable quantum information processing without the need for the nonlinear Kerr effect. This approach is called linear optical quantum computing (LOQC). Although this approach does away with the inefficiency of nonlinear processes, it requires photon counting measurements (called "post-selection") and additional photons (called "ancillary photons") to yield the desired state with some probability.

References

1. B.C. Sanders, *Quantum dynamics of the nonlinear rotator and the effects of continual spin measurement*, Phys. Rev. A 40 (1989) 2417.
2. J. Jacobson et al., *Photonic de Broglie waves*, Phys. Rev. Lett. 74 (1995) 4835.
3. M. D'Angelo, *Two-Photon Diffraction and Quantum Lithography*, Phys. Rev. Lett. 87 (2001) 013602.
4. R.W. Boyd and J.P. Dowling, *Quantum lithography, status of the field*, Quantum Information Processing 11 (2012) 891.
5. V. Giovannetti et al., *Quantum-enhanced measurements: Beating the standard quantum limit*, Science 306 (2004) 1330.
6. C.K. Hong, Z.Y. Ou and L. Mandel, *Measurement of subpicosecond time intervals between two photons by interference*, Phys. Rev. Lett. 59 (1987) 2044.
7. Frédéric Bouchard et al., *Two-photon interference: the Hong–Ou–Mandel effect*, Rep. Prog. Phys. 84 (2021) 012402.
8. C.C. Gerry and R.A. Campos, *Generation of maximally entangled photonic states with a quantum-optical Fredkin gate*, Phys. Rev. A 64 (2001) 063814.
9. Q.A. Turchette et al., *Measurement of conditional phase-shifts for quantum logic*, Phys. Rev. Lett. 75 (1995) 4710.
10. M.D. Lukin, *Trapping and manipulating photon states in atomic ensembles*, Rev. Mod. Phys. 75 (2003) 457.
11. E. Knill, R. Laflamme and G.J. Milburn, *A scheme for efficient quantum computation with linear optics*, Nature 409 (2001) 46.

Chapter 18
Quantum Imaging

Quantum imaging, involving entangled light and coincidence measurements, is providing a revolution in metrology. Nonlocal interference, ghost imaging, quantum illumination, absolute detector calibration, and interaction-free measurement are presented as leading applications. These applications demonstrate the spooky behavior of quantum mechanics. For example, we will show you how to observe an object without any photon ever interacting with it!

18.1 Nonlocal Interference

Figure 18.1 demonstrates the nonlocality of quantum mechanics in a double-slit interference experiment. A barium borate crystal ($Ba(BO_2)_2$, called BBO) is used to create entangled photon pairs by parametric down-conversion of a laser source. The two photons pass through a slit (A_1 and A_2) and a light pattern is measured by vertically scanning the detectors D_1 and D_2 (the scan direction is out of the page in Fig. 18.1, which is perpendicular to the slits). Coincidence counts (double photon detection at D_1 and D_2) are also measured for different detector positions. Single photon detection at either D_1 or D_2 results in the corresponding slit pattern, A_1 or A_2, respectively, in Fig. 18.2b. However, when coincidence counts are measured at both D_1 and D_2, then an interference pattern (Fig. 18.2c, d) results from the combined slit pattern, A_1A_2, corresponding to a double-slit. The interference pattern cannot be measured by photon counting at the individual detectors but is contained nonlocally in the entangled photon pair. The double-slit interference pattern can only be observed in the coincidence counting, even though the individual detector measurements have nothing to do with double-slit patterns. Neither beam alone contains the information required to construct the double-slit image. This experiment demonstrates "nonlocal interference".

R. LaPierre, *Getting Started in Quantum Optics*, Undergraduate Texts in Physics,
https://doi.org/10.1007/978-3-031-12432-7_18

Fig. 18.1 (a) Experiment demonstrating nonlocal interference. (b) The slit patterns, A_1 and A_2, and the combined slit pattern, A_1A_2. (Reprinted with permission from Fonseca et al. [1]. Copyright 1992 by the American Physical Society)

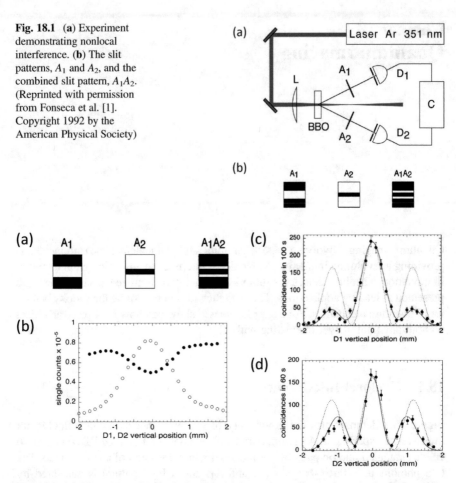

Fig. 18.2 (a) Slit patterns. (b) Single detector counts corresponding to A_1 or A_2. (c, d) D_1 and D_2 coincidence counts showing interference from the combined slit pattern, A_1A_2. (Reprinted with permission from Fonseca et al. [1]. Copyright 1992 by the American Physical Society)

18.2   Ghost Imaging

Ghost imaging (GI) (also called coincidence imaging, two-photon imaging, or correlated-photon imaging) is a technique that uses two correlated optical fields to form an image of an object. A typical GI setup (Fig. 18.3) comprises an entangled photon pair, in which one photon interacts with an object and is subsequently detected by a nonspatially resolving ("bucket") detector. The other photon of the entangled photon pair is detected by a spatially resolving detector (e.g., a 2D camera). Coincidence detection of photons at the bucket detector and photons at the imaging detector is then performed to build an image of the object placed along the bucket detector path. Neither beam alone contains the information required to

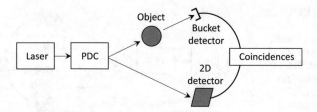

Fig. 18.3 Experiment for ghost imaging. A laser illuminates a parametric down-converter (PDC) to create entangled photon pairs. One photon is directed along an object path with a bucket detector, while the other photon is directed to a 2D detector. An image is formed from coincidence counting

reconstruct an image of the object. The bucket detector is spatially unresolved, simply detecting all photons passed by the object. Similarly, the spatially resolving detector measures the position of all photons incident on it without any information about the object. However, the correlation between these two beams can extract an image. It should be noted that GI can also be conducted using classical light (thermal source) but with lower contrast or visibility [2–4]. Entanglement provides a convenient and efficient source of correlated photons that produce greater visibility of the correlated image.

One of the advantages of GI is that it allows imaging at wavelengths for which efficient bucket detectors exist, but cameras do not. For example, one could generate pairs of entangled photons with different wavelengths (355 nm converted to 1550 nm and 460 nm). The object is illuminated at 1550 nm, but the image is formed on the 2D detector from 460 nm (in coincidence). This allows imaging in the infrared using a visible CCD camera. This method is called two-color GI [5]. GI is also of interest when imaging through diffusive media, such as free-space measurements in the presence of fog or the diffusive medium of biological samples. GI is further described in Ref. [6, 7].

18.3   Quantum Illumination

Quantum illumination achieves subshot noise imaging by coincidence counting, as shown in Fig. 18.4 [8, 9]. An object of interest (bird) is obscured by noise (cage) from thermal illumination, forming a "classical image". To discern the object from the noise, the object and a reference are illuminated with a quantum source (quantum illumination, QI). One photon of an entangled pair is measured by a spatially resolving detector forming the quantum reference, while the other photon illuminates the object, which is added to the classical image. Coincidence counting of the quantum reference and the classical image will remove the noise (cage), since the noise is not contained in the coincidence counts. Using this technique in the radio or microwave region is known as "quantum radar".

Fig. 18.4 (a) Experiment
and (b) results
demonstrating quantum
illumination (QI). (Credit:
Reprinted from Ref. [8]
under the terms of the
Creative Commons
Attribution license)

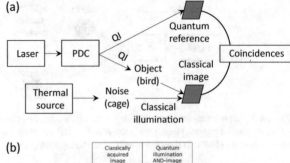

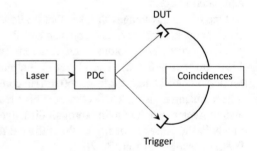

Fig. 18.5 Quantum method
of absolute photodetector
calibration

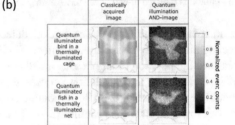

18.4   Absolute Photodetector Calibration

Quantum methods allow us to perform the calibration of photodetector efficiency
without the need for a calibrated light source [10, 11]. If we can produce a known
steady rate of single photons with known optical power on a photodetector, then we
have a means of calibrating the response of a photodetector. In Fig. 18.5, entangled
photon pairs are produced by parametric down-conversion (PDC). The total number
of trigger counts in the trigger arm, $n_{\text{trig}} = \eta_{\text{trig}} n$, is measured where η_{trig} is the
efficiency of photodetection for the trigger detector and n is the number of incident
photons. The total number of coincidence events, $n_C = \eta_{\text{DUT}} \eta_{\text{trig}} n$, is measured
where η_{DUT} is the detection efficiency of the device-under-test (DUT) in the DUT
channel. Thus, the ratio of n_C and n_{trig} gives us the unknown efficiency of the DUT:

$$\eta_{DUT} = n_C/n_{trig} \qquad (18.1)$$

The difficulty of this approach is the slow rate of single photon generation from sources available today, corresponding to only nW or less of optical power. Faster rates of single photon generation would improve absolute methods of detector calibration of use in radiometry.

18.5   Interaction-Free Measurement (Elitzur–Vaidman Bomb Problem)

One of the weirdest consequences of quantum mechanics is "interaction-free measurement" (IFM), also called the Elitzur–Vaidman (E–V) bomb problem, which was conceived in 1993 by Avshalom Elitzur and Lev Vaidman [12]. IFM is a method of detecting an object without any photon ever interacting with it [13–17] (e.g., detecting a photon-sensitive bomb without exploding it, as originally conceived by Elitzur and Vaidman).

Suppose single photons enter an interferometer, one at a time. As shown in Fig. 18.6a, the interferometer is initially adjusted so that the photon always triggers detector D_1 (with destructive interference at D_2). Now if an object is placed in the upper arm of the interferometer, as shown in Fig. 18.6b, then photons are absorbed by the object 50% of the time (50% probability), and D_1 or D_2 is triggered each with 25% probability for each photon entering the interferometer. If D_2 is triggered, then we know an object must be present. In this case, the object is detected without a photon ever having interacted with it! In fact, we can improve the detection probability to near unity by using a cavity [18]. The E–V experiment can ascertain the existence of an object in a given region of space, although no light ever "touched" the object. IFM allows imaging in the dark!

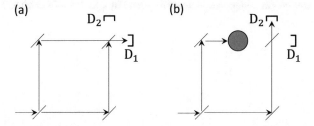

Fig. 18.6 Demonstration of interaction-free measurement. (**a**) The interferometer is set up such that, with the object absent, the probability of detection at D_1 and D_2 is 100% and 0%, respectively. (**b**) With the object present, the probability of detection at D_1 and D_2 is both 25%. In particular, detection at D_2, which was previously absent in (**a**), now occurs due to the presence of the object, although no photon interacted with the object

Exercise 18.1 Describe some applications of interaction-free measurement.

References

1. E.J.S. Fonseca et al., *Quantum interference by a nonlocal double slit*, Phys. Rev. A 60 (1992) 1530. https://doi.org/10.1103/PhysRevA.60.1530.
2. Gatti et al., *Correlated imaging, quantum and classical*, Phys. Rev. A 70 (2004) 013802.
3. R.S. Bennink et al., *Quantum and classical coincidence imaging*, Phys. Rev. Lett. 92 (2004) 033601.
4. B.I. Erkmen and J.H. Shapiro, *Signal-to-noise ratio of Gaussian-state ghost imaging*, Phys. Rev. A 79 (2009) 023833.
5. R.S. Aspden et al., *Photon-sparse microscopy: visible light imaging using infrared illumination*, Optica 2 (2015) 1049.
6. T.B. Pittman et al., *Optical imaging by means of two-photon quantum entanglement*, Phys. Rev. A 52 (1995) R3429.
7. K.W.C. Chan, M.N. O'Sullivan and R.W. Boyd, *Two-color ghost imaging*, Phys. Rev. A 79 (2009) 033808.
8. Gregory et al., *Imaging through noise with quantum illumination*, Sci. Adv. 6 (2020) eaay2652.
9. G. Brida et al., *Experimental realization of sub-shot-noise quantum imaging*, Nature Photonics 4 (2010) 227.
10. S.V. Polyakov and A.L. Migdall, *High accuracy verification of a correlated-photon-based method for determining photon-counting detection efficiency*, Opt. Express 15 (2007) 1390.
11. D.N. Klyshko, *Use of two-photon light for absolute calibration of photoelectric detectors*, Quantum Electron. 7 (1980) 1932.
12. A. Elitzur and L. Vaidman, *Quantum Mechanical Interaction-Free Measurements*, Found. Phys. 23 (1993) 987.
13. R.H. Dicke, *Interaction-free quantum measurements: A paradox?*, Am. J. Phys. 49 (1981) 925.
14. L. Vaidman, *On the realization of interaction-free measurements*, Quant. Opt. 6, 119 (1994).
15. P. Kwiat, H. Weinfurter, T. Herzog, A. Zeilinger, and M. A. Kasevich, *Interaction-free measurement*, Phys. Rev. Lett. 74 (1995) 4763.
16. A.G. White, J. R. Mitchell, O. Nairz, and P. G. Kwiat, *Interaction-free imaging*, Phys. Rev. A 58, 605 (1998).
17. Y. Zhang et al., *Interaction-free ghost-imaging of structured objects*, Optics Express 27 (2019) 2212.
18. T. Tsegaye, E. Goobar, A. Karlsson, G. Björk, M.Y. Loh and K.H. Lim, *Efficient interaction-free measurements in a high-finesse interferometer*, Phys. Rev. A 57 (1998) 3987.

Chapter 19
Light–Matter Interaction

The interaction between light and matter is essential for the generation, manipulation, and detection of quantum states used in quantum information processing. In this chapter, we describe the quantum treatment of light–matter interaction. The Jaynes–Cummings Hamiltonian is derived and used to explain spontaneous emission and Rabi oscillations. Cavity quantum electrodynamics and Rydberg atoms are introduced as a means of enhancing light–matter interaction. Collapse and revival of Rabi oscillations is examined for coherent light in a cavity. The dressed atom–cavity states are derived and used to explain vacuum Rabi splitting.

19.1 Jaynes–Cummings Hamiltonian

When dealing with the interaction of light with matter, a semiclassical description is often employed where the atom is quantized but light is treated as a classical wave (Fig. 19.1a). Here, we introduce the fully quantum approach where both the atom and light are quantized (Fig. 19.1b). We assume the atom has two states, a ground state labelled $|0\rangle$ and an excited state labelled $|1\rangle$, with energy separation $\hbar\omega_{10}$. These two states are orthonormal: $\langle i|j\rangle = \delta_{ij}$ with $i, j = 0$ or 1. The light field is quantized with orthonormal number states $|n\rangle$, and photon energy $\hbar\omega$ separating the number states.

The Hamiltonian of the combined atom–light system is

$$\widehat{H} = \widehat{H}_A + \widehat{H}_R + \widehat{H}_I \tag{19.1}$$

where $\widehat{H}_A$ is the atomic Hamiltonian, $\widehat{H}_R$ is the light (radiation) Hamiltonian, and $\widehat{H}_I$ is an interaction Hamiltonian describing the interaction between the light and the atom.

The atomic Hamiltonian can be written as

© The Author(s), under exclusive license to Springer Nature Switzerland AG 2022
R. LaPierre, *Getting Started in Quantum Optics*, Undergraduate Texts in Physics,
https://doi.org/10.1007/978-3-031-12432-7_19

Fig. 19.1 Light–matter
interaction.
(a) Semiclassical approach
where light is treated as a
classical wave and the atom
is quantized. (b) Fully
quantum approach where
both the atom and light are
quantized

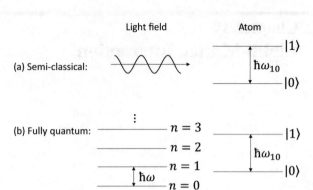

$$\widehat{H}_A = \hbar\omega_{10}|1\rangle\langle 1| \tag{19.2}$$

where the reference energy is the ground state ($E_0 = 0$). If we apply $\widehat{H}_A$ to the atomic
ground state, we get $\widehat{H}_A|0\rangle = \hbar\omega_{10}|1\rangle\langle 1|0\rangle = 0$, since $\langle 1|0\rangle = 0$. If we apply $\widehat{H}_A$ to
the excited state, we get $\widehat{H}_A|1\rangle = \hbar\omega_{10}|1\rangle\langle 1|1\rangle = \hbar\omega_{10}|1\rangle$, since $\langle 1|1\rangle = 1$. Thus,
$\widehat{H}_A$ has eigenstates $|0\rangle$ and $|1\rangle$ and energy eigenvalues 0 and $\hbar\omega_{10}$, respectively.
Equation (19.2) can also be written as

$$\widehat{H}_A = \hbar\omega_{10}\widehat{\sigma}^\dagger\widehat{\sigma} \tag{19.3}$$

where

$$\widehat{\sigma}^\dagger = |1\rangle\langle 0| \tag{19.4}$$

and

$$\widehat{\sigma} = |0\rangle\langle 1| \tag{19.5}$$

Note that $\widehat{\sigma}^\dagger\widehat{\sigma} = |1\rangle\langle 0|0\rangle\langle 1| = |1\rangle\langle 1|$ since $\langle 0|0\rangle = 1$, so we obtain Eq. (19.2)
from (19.3). $\widehat{\sigma}^\dagger$ is a raising operator because $\widehat{\sigma}^\dagger|0\rangle = |1\rangle\langle 0|0\rangle = |1\rangle$, thus raising the
atomic state from $|0\rangle$ to $|1\rangle$. $\widehat{\sigma}$ is a lowering operator because $\widehat{\sigma}|1\rangle = |0\rangle\langle 1|1\rangle = |0\rangle$,
thus lowering the atomic state from $|1\rangle$ to $|0\rangle$. Consequently, $\widehat{\sigma}^\dagger$ and $\widehat{\sigma}$ are called
raising and lowering operators, analogous to $\widehat{a}^\dagger$ and $\widehat{a}$ for the radiation. Note that
$\widehat{\sigma}^\dagger|1\rangle = |1\rangle\langle 0|1\rangle = 0$ because there is no state above $|1\rangle$, and $\widehat{\sigma}|0\rangle = |0\rangle\langle 1|0\rangle = 0$
because there is no state below $|0\rangle$; that is, in the two-level atomic system, there are
no states above the excited state nor below the ground state.

The radiation Hamiltonian is the familiar form for a quantum harmonic oscillator:

$$\widehat{H}_R = \hbar\omega\widehat{a}^\dagger\widehat{a} \tag{19.6}$$

where the vacuum energy ($\frac{1}{2}\hbar\omega$) is set equal to zero by assuming we have a lot of field quanta so that the vacuum energy is comparatively negligible.

Using the classical expression for the dipole energy, we can write the interaction Hamiltonian as

$$H_I = -\hat{\boldsymbol{p}} \cdot \widehat{\boldsymbol{E}}(\boldsymbol{r}) \tag{19.7}$$

where $\hat{\boldsymbol{p}}$ is the atomic dipole moment operator and $\widehat{\boldsymbol{E}}(\boldsymbol{r})$ is the electric field operator for the light. Quantum mechanically, the dipole moment operator can be written as

$$\hat{\boldsymbol{p}} = \boldsymbol{p}\left(|1\rangle\langle 0| + |0\rangle\langle 1|\right) \tag{19.8}$$

We see that $\hat{\boldsymbol{p}}$ couples the $|0\rangle$ and $|1\rangle$ state. Using Eqs. (19.4) and (19.5), we can write $\hat{\boldsymbol{p}}$ in the form:

$$\hat{\boldsymbol{p}} = \boldsymbol{p}\left(\hat{\sigma}^\dagger + \hat{\sigma}\right) \tag{19.9}$$

By choosing a position such that $\boldsymbol{k} \cdot \boldsymbol{r} = \frac{\pi}{2}$ in Eq. (3.44), the electric field operator can be written in a convenient form:

$$\widehat{\boldsymbol{E}} = -\boldsymbol{\varepsilon}\varepsilon^1\left(\hat{a} + \hat{a}^\dagger\right) \tag{19.10}$$

where ε^1 is the one photon amplitude:

$$\varepsilon^1 = \sqrt{\frac{\hbar\omega}{2\epsilon_o V}} \tag{19.11}$$

and $\boldsymbol{\varepsilon}$ is the electric field polarization. Thus, the interaction Hamiltonian in Eq. (19.7) can be written as

$$\widehat{H}_I = \hbar g\left(\hat{\sigma}^\dagger + \hat{\sigma}\right)\left(\hat{a} + \hat{a}^\dagger\right) \tag{19.12}$$

g describes the strength of the light–matter coupling and is given by

$$g = \frac{\boldsymbol{p} \cdot \boldsymbol{\varepsilon}\,\varepsilon^1}{\hbar} = \sqrt{\frac{\omega}{2\hbar\epsilon_o V}}\,\boldsymbol{p} \cdot \boldsymbol{\varepsilon} \tag{19.13}$$

Equation (19.12) contains four terms. The $\hat{\sigma}^\dagger\hat{a}$ term describes the absorption (annihilation) of a photon and creation of an atomic excitation. The $\hat{\sigma}^\dagger\hat{a}^\dagger$ term describes the creation of an atomic excitation and creation (emission) of a photon, which violates conservation of energy. The $\hat{\sigma}\hat{a}$ term describes the destruction of an atomic excitation and annihilation (absorption) of a photon, which also violates the conservation of energy. Finally, the $\widehat{\sigma}\widehat{a}^\dagger$ term describes the destruction of an atomic excitation and emission of a photon. The $\widehat{\sigma}^\dagger\widehat{a}^\dagger$ and $\widehat{\sigma}\widehat{a}$ terms are eliminated, since they violate the conservation of energy. Thus, only the $\widehat{\sigma}^\dagger\widehat{a}$ and $\widehat{\sigma}\widehat{a}^\dagger$ terms are kept, giving:

$$\widehat{H}_I = \hbar g\left(\widehat{\sigma}^\dagger\widehat{a} + \widehat{\sigma}\widehat{a}^\dagger\right) \tag{19.14}$$

The operators have a time dependence given by

$$\widehat{a}(t) = \widehat{a}(0)e^{-i\omega t} \tag{19.15}$$

$$\widehat{a}^\dagger(t) = \widehat{a}^\dagger(0)e^{i\omega t} \tag{19.16}$$

$$\widehat{\sigma}(t) = \widehat{\sigma}(0)e^{-i\omega_{10} t} \tag{19.17}$$

$$\widehat{\sigma}^\dagger(t) = \widehat{\sigma}^\dagger(0)e^{i\omega_{10} t} \tag{19.18}$$

Thus, in the Heisenberg picture, the interaction Hamiltonian from Eq. (19.12) takes the form:

$$\widehat{H}_I = \hbar g\left(\widehat{\sigma}^\dagger\widehat{a}e^{i(\omega_{10}-\omega)t} + \widehat{\sigma}^\dagger\widehat{a}^\dagger e^{i(\omega_{10}+\omega)t} + \widehat{\sigma}\widehat{a}e^{-i(\omega_{10}+\omega)t} + \widehat{\sigma}\widehat{a}^\dagger e^{-i(\omega_{10}-\omega)t}\right) \tag{19.19}$$

The $\widehat{\sigma}^\dagger\widehat{a}^\dagger$ term and the $\widehat{\sigma}\widehat{a}$ term vary much more rapidly compared to the other two terms. Thus, the fast evolution of the $\widehat{\sigma}^\dagger\widehat{a}^\dagger$ and $\widehat{\sigma}\widehat{a}$ terms averages to zero during the much slower time evolution of the other two terms. Ignoring the $\widehat{\sigma}^\dagger\widehat{a}^\dagger$ and $\widehat{\sigma}\widehat{a}$ terms in this manner is called the "rotating wave approximation (RWA)". This provides another justification for ignoring these terms, besides being unphysical (violating conservation of energy) as described above.

Combining Eqs. (19.2), (19.6) and (19.14) into Eq. (19.1) gives

$$\widehat{H} = \hbar\omega_{10}\widehat{\sigma}^\dagger\widehat{\sigma} + \hbar\omega\widehat{a}^\dagger\widehat{a} + \hbar g\left(\widehat{\sigma}^\dagger\widehat{a} + \widehat{\sigma}\widehat{a}^\dagger\right) \tag{19.20}$$

Equation (19.20) is known as the Jaynes–Cummings Hamiltonian, introduced by E.T. Jaynes and F.W. Cummings in the 1960s [1]. Although relatively simple, the Jaynes–Cummings Hamiltonian predicts a wide range of phenomenon.

19.2 Spontaneous Emission

The light–matter system in Fig. 19.1 can be described by a quantum state $|i, n\rangle$ where $i = 0$ or 1 indicates the atomic state, and $n = 0, 1, 2, \ldots$ indicates the Fock (number) state of the light. Suppose the initial state is $|0, n\rangle$, meaning the atom occupies the ground state, and the light field contains n photons in the Fock state. The absorption of a photon results in the final state $|1, n - 1\rangle$; that is, the atom is promoted to the excited state and the field loses one photon by annihilation (absorption). The dipole coupling between the initial and final states is given by

$$\langle 1, \ n - 1 | \widehat{H}_I | 0, \ n \rangle = \langle 1, \ n - 1 | \hbar g (\widehat{\sigma}^\dagger \widehat{a} + \widehat{\sigma} \widehat{a}^\dagger) | 0, \ n \rangle \tag{19.21}$$

$$= \hbar g \sqrt{n} \tag{19.22}$$

Conversely, the emission of a photon from an initial state $|1, n\rangle$ results in the final state $|0, n + 1\rangle$. The dipole coupling between the initial and final states in this case is

$$\langle 0, \ n + 1 | \widehat{H}_I | 1, \ n \rangle = \langle 0, \ n + 1 | \hbar g \left(\widehat{\sigma}^\dagger \widehat{a} + \widehat{\sigma} \widehat{a}^\dagger \right) | 1, \ n \rangle \tag{19.23}$$

$$= \hbar g \sqrt{n + 1} \tag{19.24}$$

Exercise 19.1 Derive Eqs. (19.22) and (19.24).

Equation (19.24) indicates that we can have emission of a photon from the excited atom even in the case of $n = 0$; that is, the initial state is vacuum. This explains the process of spontaneous emission from an atom in an excited state even in the absence of any externally applied field. We can think of the field fluctuations in the vacuum as causing spontaneous emission.

19.3   Cavity Quantum Electrodynamics

Equation (19.14) indicates that strong light–matter interactions can occur when the light–atom coupling, g, is large. Equation (19.13) shows that one method of making g large is to make the mode volume V small. This can be done by trapping the atom in a cavity of small volume formed by two mirrors, as shown in Fig. 19.2 (methods of atom trapping are discussed in Chap. 21). This approach is known as cavity quantum electrodynamics (CQED). In this case, we can think of the atom-photon interaction as being enhanced by the many round trips that a photon makes across the atom due to reflection back and forth between the mirrors [2]. There are many other ways of implementing CQED such as micropillars, microdisks, microspheres, and photonic crystals, as shown in Fig. 19.3 [3]. We have already seen some other consequences of CQED such as the Purcell effect in Chap. 4.

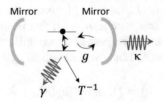

Fig. 19.2 Illustration of cavity quantum electrodynamics (CQED). γ, T^{-1} and κ denote various loss mechanisms (γ: rate of spontaneous emission into free space; κ: rate of photon transmission due to finite reflectivity of the cavity mirrors; T^{-1}: rate of atom escape from the cavity)

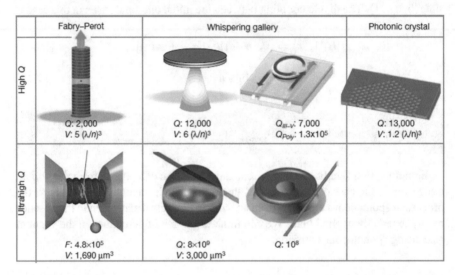

Fig. 19.3 Illustration of various CQED implementations. The microcavities are organized by column according to the confinement method used and by row according to quality factor Q of the cavity. Upper row: micropost or pillar, microdisk, add/drop filter, photonic crystal cavity. Lower row: Fabry-Perot bulk optical cavity, microsphere, and microtoroid. n is the material refractive index, V is the mode volume, Q is the quality factor, and F is the finesse. Two Q values are cited for the add/drop filter: one for a polymer design, Q_{Poly}, and the second for a III–V semiconductor design, Q_{III-V}. (Reprinted by permission from Springer Nature, Vahala [3]. Copyright 2003)

Exercise 19.2 Explore the microcavities in Fig. 19.3 and explain how the devices are fabricated.

19.4 Circuit QED

Another way to enhance atom–light interaction is to create "artificial atoms", a system that behaves like a two-level atom but with a stronger coupling. A common "artificial atom" is a superconducting circuit containing Josephson junction circuit elements that act as a two-level system. The superconducting circuit (Fig. 19.4c) is coupled to a microwave resonator (Fig. 19.4a, c). Due to the small mode volumes made possible by the microwave waveguide, the g coupling between microwave photons and the superconducting circuit can be orders of magnitude larger than real atomic systems. These superconducting circuits are called "circuit QED" (cQED) and are presently among the leading technologies in quantum computing.

19.5 Rydberg Atoms

Another method to increase the atom–light coupling, g, according to Eq. (19.13), is to increase the dipole moment p of the atom in the cavity of a CQED system. Rydberg atoms are atoms with a valence electron excited to a high principal quantum number ($n\sim50$) with a large orbital radius. In such an orbit, the balance between the

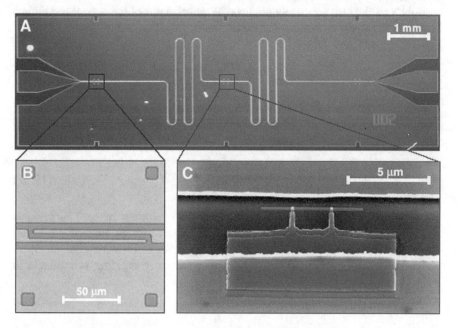

Fig. 19.4 (a) Optical micrograph showing circuit quantum electrodynamics (cQED). The transmission line "wiggles" to increase its cavity length. (b) The cavity mirror is formed by a gap in the transmission line. (c) The superconducting circuit (called a "transmon qubit") with two Josephson junctions. (Reprinted by permission from Springer Nature, Wallraff et al. [4]. Copyright 2004)

Fig. 19.5 A Rydberg atom resembles a hydrogen atom (not to scale!)

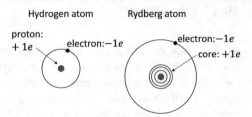

charge of the inner electrons and the nucleus produces a hydrogen-like system with charge $+e$ surrounded by a single valence electron. Thus, the valence electron feels a hydrogen-like Coulomb potential (Fig. 19.5). According to Bohr's simple model of the hydrogen atom, familiar from introductory quantum mechanics, the radius of the electron orbit is given by

$$r = n^2 a_0 \qquad (19.25)$$

where $a_0 \sim 0.53$ Å is the Bohr radius and n is the principal quantum number. According to Eq. (19.25), the average radius of the electron orbit scales as n^2, and the resulting electron orbit is about 2500 atomic diameters for $n \sim 50$; that is, on the order of 0.1 micron! This results in a huge dipole moment ($p = er$) due to the large electron-nucleus separation (r).

Bohr's model also accurately predicts the wavelengths of the energy transitions, known as the Rydberg formula (hence the name, "Rydberg atoms"):

$$\frac{1}{\lambda} = R_{\mathrm{H}} \left(\frac{1}{n_1{}^2} - \frac{1}{n_2{}^2} \right) \qquad (19.26)$$

where $R_{\mathrm{H}} \sim 1.0974 \times 10^7$ m^{-1} is the Rydberg constant for hydrogen, and n_1 and n_2 are the principal quantum numbers associated with the transition. For large n and for transitions between adjacent energy levels, we can approximate Eq. (19.26) as

$$\frac{1}{\lambda} = R_{\mathrm{H}} \left(\frac{2}{n^3} \right) \qquad (19.27)$$

For $n \sim 50$, Eq. (19.27) gives $\lambda \sim 6$ mm, that is, in the microwave range. This gives us an idea of the required cavity dimensions to support a single mode field. Using Rydberg states, interactions between atoms are more easily controlled [5, 6]. A review of Rydberg atoms in the context of quantum computing is available in Ref. [7].

19.6 Rabi Oscillations

Suppose the initial state of an atom–light system in a cavity is $|1, n\rangle$, meaning we start in the excited state of the atom ($|1\rangle$) with n photons in the cavity ($|n\rangle$). Following an atomic transition to the ground state accompanied by photon emission, the state becomes $|0, n+1\rangle$; that is, we add one photon to the cavity. The atom–light system can also exist in a superposition of these two states:

$$|\psi(t)\rangle = c_{1,n}(t)|1,\ n\rangle + c_{0,n+1}(t)|0,\ n+1\rangle \qquad (19.28)$$

In general, Eq. (19.28) represents an entangled state of the combined atom-field system.

The dynamics of the system are given by the time-dependent Schrodinger equation:

$$i\hbar\frac{d|\psi(t)\rangle}{dt} = \widehat{H}|\psi(t)\rangle \qquad (19.29)$$

The right-hand side of Eq. (19.29) is evaluated using the Jaynes–Cummings Hamiltonian:

$$\widehat{H}|\psi(t)\rangle = \left[\hbar\omega_{10}\widehat{\sigma}^{\dagger}\widehat{\sigma} + \hbar\omega\widehat{a}^{\dagger}\widehat{a} + \hbar g\left(\widehat{\sigma}^{\dagger}\widehat{a} + \widehat{\sigma}\widehat{a}^{\dagger}\right)\right]$$
$$\times (c_{1,n}|1,\ n\rangle + c_{0,n+1}|0,\ n+1\rangle) \qquad (19.30)$$

Using Eq. (2.127) and (2.128), we get

$$\widehat{H}|\psi(t)\rangle = \hbar\omega_{10}c_{1,n}|1,\ n\rangle + \hbar\omega n c_{1,n}|1,\ n\rangle + \hbar\omega(n+1)c_{0,n+1}|0,\ n+1\rangle$$
$$+ \hbar g c_{0,n+1}\sqrt{n+1}\,|1,\ n\rangle + \hbar g c_{1,n}\sqrt{n+1}\,|0,\ n+1\rangle \qquad (19.31)$$
$$= E_i c_{1,n}|1,\ n\rangle + E_f c_{0,n+1}|0,\ n+1\rangle + \hbar g c_{0,n+1}\sqrt{n+1}\,|1,\ n\rangle$$
$$+ \hbar g c_{1,n}\sqrt{n+1}\,|0,\ n+1\rangle \qquad (19.32)$$

where $E_i = \hbar\omega_{10} + \hbar\omega n$ is the initial energy with the atom in the excited state and n photons in the field, and $E_f = \hbar\omega(n+1)$ is the final energy with the atom in the ground state and $n+1$ photons in the field. Here, we assume $\hbar\omega_{10} = \hbar\omega$; that is, the cavity field is resonant with the atomic transition. The left-hand side of Eq. (19.29) is

$$i\hbar\frac{d|\psi(t)\rangle}{dt} = i\hbar\dot{c}_{1,n}|1, n\rangle + i\hbar c_{1,n}\frac{d|1, n\rangle}{dt} + i\hbar\dot{c}_{0,n+1}|0, n+1\rangle$$

$$+ i\hbar c_{0,n+1}\frac{d|0, n+1\rangle}{dt} \tag{19.33}$$

$$= i\hbar\dot{c}_{1,n}|1, n\rangle + E_i c_{1,n}|1, n\rangle + i\hbar\dot{c}_{0,n+1}|0, n+1\rangle + E_f c_{0,n+1}|0, n+1\rangle \tag{19.34}$$

Equating Eqs. (19.32) and (19.34) gives

$$\dot{c}_{1,n} = -ig\sqrt{n+1}\, c_{0,n+1} \tag{19.35}$$

and

$$\dot{c}_{0,n+1} = -ig\sqrt{n+1}\, c_{1,n} \tag{19.36}$$

Equations (19.35) and (19.36) are two coupled differential equations. Combining these equations gives

$$\ddot{c}_{1,n} = -g^2(n+1)c_{1,n} \tag{19.37}$$

The initial condition is $|1, n\rangle$; that is, the atom is initially in the excited state. Thus, $c_{1,n}(0) = 1$ and $c_{0,n+1}(0) = 0$. Solving Eq. (19.37) with this initial condition gives

$$c_{1,n}(t) = \cos\left(gt\sqrt{n+1}\right) \tag{19.38}$$

and substituting Eq. (19.38) into (19.36) and solving gives

$$c_{0,n+1}(t) = -i\sin\left(gt\sqrt{n+1}\right) \tag{19.39}$$

Finally, substituting Eqs. (19.38) and (19.39) into Eq. (19.28) gives

$$|\psi(t)\rangle = \cos\left(\frac{\Omega_n t}{2}\right)|1, n\rangle - i\sin\left(\frac{\Omega_n t}{2}\right)|0, n+1\rangle \tag{19.40}$$

where

$$\Omega_n = 2g\sqrt{n+1} \tag{19.41}$$

Ω_n is called the Rabi frequency, named after Isidor I. Rabi (Fig. 19.6).

In the semiclassical model, where the atom is quantized but the light field is treated as a classical field, the Rabi frequency is $\Omega = p \cdot E/\hbar$ where $p \cdot E$ is the classical expression for the dipole energy. In the fully quantum model, where both the atom and the field are quantized, the Rabi frequency is quantized according to Eq. (19.41).

Fig. 19.6 Isidor I. Rabi
(Nobel Prize in Physics in
1944). (Credit: Wikimedia
Commons [8])

Fig. 19.7 Probabilities
$P_{1,n}(t)$ (dashed line) and
$P_{0,n+1}(t)$ (solid line),
showing Rabi oscillations

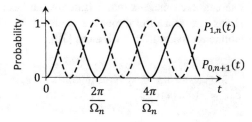

According to Eq. (19.40), the probability of the atom-field system being in the
$|1, n\rangle$ state is

$$P_{1,n}(t) = |c_{1,n}|^2 = \cos^2\left(\frac{\Omega_n t}{2}\right) \tag{19.42}$$

and the probability of being in the $|0, n+1\rangle$ state is

$$P_{0,n+1}(t) = |c_{0,n+1}|^2 = \sin^2\left(\frac{\Omega_n t}{2}\right) \tag{19.43}$$

We see that the state oscillates between the $|1, n\rangle$ and $|0, n+1\rangle$ state with the Rabi
frequency Ω_n. These are called Rabi oscillations as shown in Fig. 19.7.

Equation (19.41) shows that Rabi oscillations exist even in the initial absence of
light ($n = 0$) with frequency $\Omega_0 = 2g$. These are called vacuum Rabi oscillations,
caused by spontaneous emission of the atom by the vacuum. The emitted photon is
given up to the cavity where it can be repeatedly absorbed and remitted by the atom.
The observation of vacuum Rabi oscillations was demonstrated in Ref. [9].

Of course, no system is perfect, and the Rabi oscillations eventually decay due to
losses and decoherence associated with coupling to the environment (Fig. 19.2). By
making the light–atom coupling (g) large by using a cavity, many Rabi oscillations
are possible before photon loss due to spontaneous emission into free space (rate γ),

photon transmission loss due to finite reflectivity of the cavity mirrors (rate κ), and atom escape from the cavity (rate T^{-1}), as shown in Fig. 19.2. CQED aims to have $g \gg \kappa, \gamma, T^{-1}$ so that many cycles of the Rabi oscillations can be observed before decay. This is the so-called strong coupling regime of CQED where light and atomic quanta play a dominant role in the system dynamics. Often, one characterizes the system by combining the parameters into a single dimensionless quantity called the cooperativity, defined as (assuming T^{-1} is negligible):

$$C = \frac{g^2}{\kappa \gamma} \tag{19.44}$$

Obviously, one wants $C \gg 1$ for the strong coupling regime. In the strong coupling regime, irreversible spontaneous emission changes to a reversible exchange of energy between the atom and the cavity mode. With sufficiently strong coupling, one can even make a maser (microwave laser), for example, by passing Rydberg atoms one at a time through a cavity, where field buildup occurs by cumulative atomic emission from each atom [10].

Exercise 19.3 Give typical values for g, κ, γ, and T^{-1} for a few cavity systems (e.g., from Fig. 9.3).

19.7   Collapse and Revival of Rabi Oscillations

Let us suppose that a coherent state exists inside a cavity with a two-level atom. Recall that the coherent state in the number representation is

$$|\alpha\rangle = \sum_{n=0}^{\infty} c_n |n\rangle \tag{19.45}$$

with Poisson probability distribution of the photon number n:

$$|c_n|^2 = e^{-\langle n \rangle} \frac{\langle n \rangle^n}{n!} \tag{19.46}$$

In the coherent state, each of the Fock states n will undergo Rabi oscillation with frequency given by Eq. (19.41). Let's see what happens when we superimpose all these Rabi oscillations.

We assume the atom is initially in a superposition of the ground and excited state:

$$|\psi(0)\rangle_{\text{atom}} = c_0 |0\rangle + c_1 |1\rangle \tag{19.47}$$

Also, we assume the field is initially in a coherent state according to Eq. (19.45):

$$|\psi(0)\rangle_{\text{field}} = \sum_{n=0}^{\infty} c_n |n\rangle \tag{19.48}$$

The initial atom-field state is then given by the tensor product:

$$|\psi(0)\rangle = |\psi(0)\rangle_{\text{atom}} \otimes |\psi(0)\rangle_{\text{field}} \tag{19.49}$$

If we take the case of $c_0 = 0$ and $c_1 = 1$ (the atom is initially in the excited state), then the solution to Schrodinger's equation (similar to Eqs. (19.38) and (19.39)) gives

$$|\psi(t)\rangle = \sum_{n=0}^{\infty} c_n \left[\cos\left(gt\sqrt{n+1}\right)|1, \ n\rangle - i\sin\left(gt\sqrt{n+1}\right)|0, \ n+1\rangle \right] \tag{19.50}$$

We define the atomic inversion, which is the difference in probabilities of the atom in the excited and ground state:

$$W(t) = P_1(t) - P_0(t) \tag{19.51}$$

$$= \sum_{n=0}^{\infty} |c_n|^2 \cos\left(2gt\sqrt{n+1}\right) \tag{19.52}$$

$$= e^{-\langle n\rangle} \sum_{n=0}^{\infty} \frac{\langle n\rangle^n}{n!} \cos\left(2gt\sqrt{n+1}\right) \tag{19.53}$$

Equation (19.53) is plotted in Fig. 19.8, showing the Rabi oscillations. For large coherent fields, the Rabi frequency is $2g\sqrt{\langle n\rangle + 1}$, associated with the mean photon number $\langle n\rangle$, since the dispersion $\Delta n = \sqrt{\langle n\rangle}$ is small. However, the Rabi oscillations eventually decay (collapse), followed by a revival, followed by another collapse, etc. This is known as collapse and revival of the Rabi oscillations. This occurs due to the interference of Rabi oscillations with different quantized Rabi frequencies. The Rabi oscillations fall in and out of phase, leading to destructive and

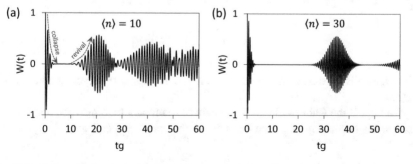

Fig. 19.8 Collapse and revival of Rabi oscillations for average photon number of (**a**) $\langle n\rangle = 10$ and (**b**) $\langle n\rangle = 30$. Time t is in units of $1/g$

constructive interference. In other words, a beat note is formed between the different Rabi frequencies.

Note that Rabi oscillations can be explained by a semiclassical theory where the atomic system has quantized energy levels but the light field is classical. However, the repeated collapse and revival of the Rabi oscillations can only be explained with quantized fields via the Jaynes–Cummings model. This is the reason why collapse and revival of the Rabi oscillations are of such great interest in quantum optics. The first observation of Rabi collapse and revival is Ref. [11].

The collapse of the Rabi oscillations occurs when all the different Fock state components become out of phase by π and destructively interfere. For a coherent field with large average photon number $\langle n \rangle$, we can approximate the photon distribution as a Gaussian with a standard deviation $\sqrt{\langle n \rangle}$. Thus, we can estimate the collapse time t_C as occurring when

$$\Omega_{\langle n \rangle + \Delta n} t_C - \Omega_{\langle n \rangle - \Delta n} t_C = \pi \tag{19.54}$$

According to Eq. (19.54), we get

$$2g\sqrt{\langle n \rangle + \Delta n}\, t_C - 2g\sqrt{\langle n \rangle - \Delta n}\, t_C = \pi \tag{19.55}$$

or

$$2g\sqrt{\langle n \rangle}\sqrt{1 + \frac{1}{\sqrt{\langle n \rangle}}}\, t_C - 2g\sqrt{\langle n \rangle}\sqrt{1 - \frac{1}{\sqrt{\langle n \rangle}}}\, t_C = \pi \tag{19.56}$$

Since $1/\sqrt{\langle n \rangle} \ll 1$ for large $\langle n \rangle$, we can use a binomial expansion in Eq. (19.56), which gives

$$2g\sqrt{\langle n \rangle}\left(1 + \frac{1}{2\sqrt{\langle n \rangle}}\right) t_C - 2g\sqrt{\langle n \rangle}\left(1 - \frac{1}{2\sqrt{\langle n \rangle}}\right) t_C = \pi \tag{19.57}$$

or

$$t_C = \frac{\pi}{2g} \tag{19.58}$$

Thus, the collapse time is independent of photon number as observed qualitatively in Fig. 19.8.

We expect the Rabi oscillations to begin reviving at a time t_R when neighboring Fock states become in phase giving constructive interference. Thus,

$$\Omega_{\langle n \rangle + 1} t_R - \Omega_{\langle n \rangle} t_R = 2\pi \tag{19.59}$$

$$2g\sqrt{\langle n \rangle + 1}\, t_R - 2g\sqrt{\langle n \rangle}\, t_R = 2\pi \tag{19.60}$$

$$2g\sqrt{\langle n \rangle}\sqrt{1 + \frac{1}{\langle n \rangle}}\, t_R - 2g\sqrt{\langle n \rangle}\, t_R = 2\pi \tag{19.61}$$

Using a binomial expansion, we get

$$2g\sqrt{\langle n \rangle}\left(1 + \frac{1}{2\langle n \rangle}\right) t_R - 2g\sqrt{\langle n \rangle}\, t_R = 2\pi \tag{19.62}$$

or

$$t_R = \frac{2\pi\sqrt{\langle n \rangle}}{g} \tag{19.63}$$

Thus, the revival time is related to the average photon number, as observed qualitatively in Fig. 19.8.

19.8   Example of a CQED Experiment

Figure 19.9 illustrates a cavity QED experiment [12]. An oven (O) produces a beam of rubidium atoms which are then prepared into Rydberg states (B). The Rydberg atoms enter a cavity (C) containing a coherent cavity mode produced by a microwave generator (S). While in the cavity, the Rydberg atoms undergo Rabi oscillations between two principal quantum numbers ($n = 50$ and 51). The Rydberg atoms exit the cavity and their state is detected. The detection mechanism uses an electric field to strip the electron in the excited state ($n = 51$) off the atom where it is then detected with a channeltron electron multiplier. If the atom is in the ground state ($n = 50$), the electric field is not sufficient to ionize the atom. Thus, the state of the atom (ground or excited state) can be determined by charge detection. The

Fig. 19.9 Experimental setup for the results of Fig. 19.10. O: oven; B: Rydberg atom state preparation; S: coherent source; C: cavity; D: detector. (Reprinted with permission from Brune et al. [12]. Copyright 1996 by the American Physical Society)

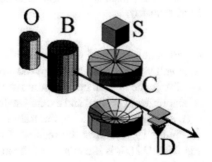

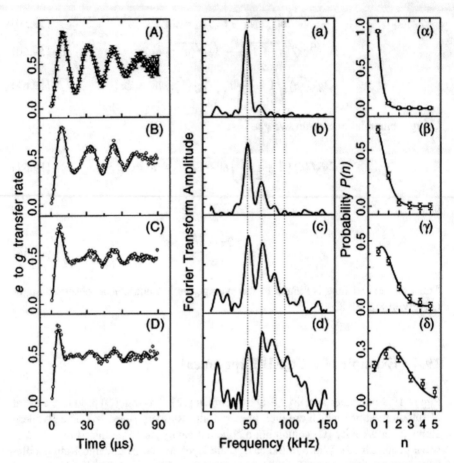

Fig. 19.10 The probability versus time of finding the Rydberg atom in the ground state, showing Rabi oscillations due to (**A**) vacuum field with $\langle n \rangle = 0.06$ (due to a small thermal field) and (**B–D**) coherent field with $\langle n \rangle = 0.40, 0.85$ and 1.77, respectively. The points are experimental (with error bars in (**A**) only for clarity); the solid lines correspond to theoretical fits. (**a–d**) Corresponding Fourier transforms. Frequencies $\nu = 47$ kHz, $\sqrt{2}\nu$, $\sqrt{3}\nu$, and $\sqrt{4}\nu$ are indicated by vertical dotted lines. (**α – δ**) Corresponding photon number distribution inferred from experimental signals (points). Solid lines show the theoretical thermal (**α**) or coherent (**β, γ, δ**) distributions, which best fit the data. (Reprinted with permission from Brune et al. [12]. Copyright 1996 by the American Physical Society)

experiment is repeated with many different travel times through the cavity due to the Maxwell velocity distribution of the atoms. The travel times are known according to the difference in time between the state preparation and detection of the atom. Thus, probabilities of ground and excited state populations can be obtained as a function of time and plotted to observe the Rabi oscillations.

Figure 19.10 shows the results of an experiment performed using the setup of Fig. 19.9 [12]. It is illustrative of the kind of results that are obtained in cavity QED experiments. Figure 19.10(A) shows vacuum Rabi oscillations with a frequency $\Omega_0 = 47$ kHz obtained from the Rb atoms passing through the cavity. In Fig. 19.10 (B–D), different microwave field amplitudes were turned on, corresponding to a coherent field with different average photon numbers in the cavity. Here, a weak collapse and revival of the oscillations can be observed. Figure 19.10(a–d) are the Fourier transforms of (A–D), showing the discrete frequency components of the Rabi oscillations and the quantized nature of the field. Figure 19.10(a) shows the case of vacuum Rabi oscillations with frequency $\Omega_0 = 47$ kHz, while Fig. 19.10(b–d) show additional frequency components when the microwave generator is turned on, corresponding to the coherent field. The Rabi frequencies of each Fock state in the coherent field are given by $\Omega_n = \Omega_0\sqrt{n+1}$, according to Eq. (19.41). Figure 19.10(b–d) shows the expected $\sqrt{n}$ nonlinearity. Figure 19.10(α–δ) shows the amplitudes of the frequency components. The amplitude components follow a thermal radiation distribution in Fig. 19.10(α) in the absence of any microwave photons, and a Poisson distribution in (β–δ) as expected for a coherent field (with the microwave source turned on). The average photon numbers are obtained from these distributions.

19.9   Dressed Atom–Cavity States and Vacuum Rabi Splitting

In Fig. 19.1b, we treated the quantized levels of the atom and of the cavity field as separate. The system is described by states of the form:

$$|\psi_1\rangle = |1,\ n\rangle \tag{19.64}$$

$$|\psi_2\rangle = |0,\ n+1\rangle \tag{19.65}$$

Using these as the basis states, we can write the Jaynes–Cummings Hamiltonian as a matrix with elements given by

$$\widehat{H}_{ij} = \left\langle \psi_i |\widehat{H}| \psi_j \right\rangle \tag{19.66}$$

where i and j are indices indicating the matrix element ($i, j = 0, 1$), which gives

$$\widehat{H} = \begin{pmatrix} \hbar\omega n + \dfrac{\hbar\omega_{10}}{2} & \hbar g\sqrt{n+1} \\[2mm] \hbar g\sqrt{n+1} & \hbar\omega(n+1) - \dfrac{\hbar\omega_{10}}{2} \end{pmatrix} \tag{19.67}$$

where we have shifted the zero of energy to $\frac{\hbar\omega_{10}}{2}$ between the atomic ground and excited state energy, and we have ignored the zero-point energy. The diagonal elements correspond to the energy eigenvalues without the interaction term (ground state energy with $n + 1$ photons in the field, and excited state energy with n photons in the field), while the off-diagonal elements correspond to the interaction term of the Hamiltonian (dipole interaction).

At resonance ($\omega = \omega_{10}$), the new eigenstates of the coupled atom–cavity system are

$$|n_\pm\rangle = \frac{1}{\sqrt{2}}(\pm|1,\ n\rangle + |0,\ n + 1\rangle) \tag{19.68}$$

or, in the vector representation,

$$|n_\pm\rangle = \frac{1}{\sqrt{2}}\begin{pmatrix} \pm 1 \\ 1 \end{pmatrix} \tag{19.69}$$

with corresponding eigenvalues:

$$E_\pm = \hbar\omega\left(n + \frac{1}{2}\right) \pm \hbar g\sqrt{n + 1} \tag{19.70}$$

$|n_\pm\rangle$ are called the dressed states (i.e., dressed by the photons), while the basis states, $|1, n\rangle$ and $|0, n + 1\rangle$, are called the "bare" states. The dressed states are entangled; that is, they cannot be expressed as a tensor product of states involving only the atom and another part involving only the field.

Exercise 19.4
Check that Eqs. (19.68) and (19.70) satisfy the time-independent Schrodinger equation, $\widehat{H}|n_\pm\rangle = E_\pm|n_\pm\rangle$.

For the off-resonance case ($\omega \neq \omega_{10}$), the solution to Schrodinger's equation gives

$$E_\pm = \hbar\omega\left(n + \frac{1}{2}\right) \pm \frac{\hbar}{2}\Omega_n(\Delta) \tag{19.71}$$

where $\Omega_n(\Delta)$ is called the generalized Rabi frequency that is defined as

$$\Omega_n(\Delta) = \sqrt{\Delta^2 + 4g^2(n + 1)} = \sqrt{\Delta^2 + \Omega_n{}^2} = \sqrt{\Delta^2 + \Omega_0{}^2(n + 1)} \tag{19.72}$$

and Δ is called the detuning that is defined as

$$\Delta = \omega_{10} - \omega \tag{19.73}$$

The eigenstates corresponding to Eq. (19.71) become

$$|n_+\rangle = \cos{(\theta/2)}|1,\ n\rangle + \sin{(\theta/2)}|0,\ n+1\rangle \tag{19.74}$$

and

$$|n_-\rangle = -\sin{(\theta/2)}|1,\ n\rangle + \cos{(\theta/2)}|0,\ n+1\rangle \tag{19.75}$$

where θ is called the mixing angle with

$$\sin{(\theta/2)} = \frac{1}{\sqrt{2}}\sqrt{\frac{\Omega_n(\Delta) - \Delta}{\Omega_n(\Delta)}} \tag{19.76}$$

$$\cos{(\theta/2)} = \frac{1}{\sqrt{2}}\sqrt{\frac{\Omega_n(\Delta) + \Delta}{\Omega_n(\Delta)}} \tag{19.77}$$

Exercise 19.5 Show that at resonance, Eqs. (19.74) and (19.75) reduce to Eq. (19.68), and Eq. (19.71) reduces to Eq. (19.70). Show that far from resonance, Eqs. (19.74) and (19.75) reduce to the bare states, $|1, n\rangle$ and $|0,\ n + 1\rangle$.

We can represent the energy levels as shown in Fig. 19.11, which is called the "Jaynes–Cummings ladder". The uncoupled energy levels are shown on the left of Fig. 19.11. At resonance, each pair of levels ($|1, n\rangle$, $|0, n + 1\rangle$) would be degenerate. Off resonance, there is a difference of energy ($\hbar\Delta$) for the $|1, n\rangle$ and $|0, n + 1\rangle$ state. Figure 19.11 shows the case of red-detuning ($\Delta = \omega_{10} - \omega > 0$). The energy levels are split by the atom-field coupling, as shown on the right of Fig. 19.11. This shift of the energy levels is called the "AC Stark effect". We will consider this effect further in Chap. 21 for its use in atom cooling.

The splitting of the energy levels can be measured by a pump-probe experiment where a strong pump laser drives the splitting of the dressed states, while a weaker probe laser drives a transition with another energy level. The observed energy doublet is known as the Autler-Townes effect and is a manifestation of the AC Stark effect.

The energy splitting near resonance is known as an "avoided crossing". This is a common phenomenon in other coupled oscillators such as two coupled pendulums or two coupled LC electronic oscillators. The similar resonance frequencies of two oscillators become split when the two oscillators are coupled together.

At resonance ($\Delta = 0$), the coupled energy levels are split by an amount $\pm\hbar g\sqrt{n+1}$. The splitting at $n = 0$ becomes $\pm\hbar g = \pm\hbar\Omega_0/2$, which is known as the vacuum Rabi splitting (VRS). VRS was first observed in Ref. [13] by measuring transmission spectra through a cavity with a single mode and one atom present on average. Since then, VRS has been observed in a wide variety of systems [14]. The observation of VRS and corresponding avoided crossings is an indication of the strong coupling regime.

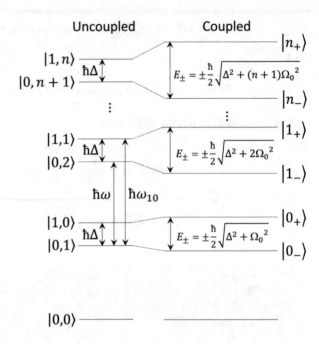

Fig. 19.11 Jaynes–Cummings ladder for red-detuned light. The left-hand side shows the energy levels for the uncoupled atom–light states, and the right-hand side shows the energy levels for the coupled atom–light states

References

1. E.T. Jaynes and F.W. Cummings, *Comparison of quantum and semiclassical radiation theories with application to the beam maser*, Proc. IEEE 51 (1963) 89.
2. G. Rempe, R. J. Thompson, H. J. Kimble and R. Lalezari, *Measurement of ultralow losses in an optical interferometer*, Optics Letters 17 (1992) 363.
3. K.J. Vahala, *Optical microcavities*, Nature 424 (2003) 839.
4. A. Wallraff et al., *Strong coupling of a single photon to a superconducting qubit using circuit quantum electrodynamics*, Nature 431 (2004) 162.
5. E. Urban et al., *Observation of Rydberg blockade between two atoms*, Nature Phys. 5 (2009) 110.
6. A. Browaeys, D. Barredo and T. Lahaye, *Experimental investigations of dipole–dipole interactions between a few Rydberg atoms*, J. Phys. B: At. Mol. Opt. Phys. 49 (2016) 152001.
7. M. Saffman, T.G. Walker and K. Mølmer, *Quantum information with Rydberg atoms*, Rev. Mod. Phys. 82 (2010) 2313.
8. File: II Rabi.jpg. (2020, October 21). *Wikimedia Commons, the free media repository*. Retrieved 15:45, December 7, 2020 from https://commons.wikimedia.org/w/index.php?title=File:II_Rabi.jpg&oldid=496554939.
9. R. Bose, T. Cai, K.R. Choudhury, G.S. Solomon and E. Waks, *All-optical coherent control of vacuum Rabi oscillations*, Nature Photon. 8 (2014) 858.
10. D. Meschede, H. Walther and G. Müller, *One-atom maser*, Phys. Rev. Lett. 54 (1985) 551.

11. G. Rempe, H. Walther and N. Klein, *Observation of quantum collapse and revival in a one-atom maser*, Phys. Rev. Lett. 58 (1987) 353.

12. M. Brune et al., *Quantum Rabi oscillation: A direct test of field quantization in a cavity*, Phys. Rev. Lett. 76 (1996) 1800.

13. R.J. Thompson, G. Rempe, and H.J. Kimble, *Observation of normal-mode splitting for an atom in an optical cavity*, Phys. Rev. Lett. 68 (1992) 1132.

14. G. Khitrova, *Vacuum Rabi splitting in semiconductors*, Nature Physics 2 (2006) 81.

Chapter 20
Atomic Clock

In the late 1930s, Isidor Rabi introduced the idea of using atomic resonances as frequency standards, that is, as a means of keeping time, which eventually became the atomic clock. The atomic clock is a perfect example of light-matter interaction being put to practical use. Due to the atomic clock, we can measure time with less uncertainty than any other physical quantity.

20.1 Quartz Oscillators—Before Atomic Time

To put the atomic clock in context, let us first describe a ubiquitous time-keeping technology that predates the atomic clock—the quartz crystal oscillator. Bell labs built the first quartz crystal oscillator clock in 1927, which soon began to replace pendulums or other mechanical clocks as the standard for time measurement. Quartz oscillators are based on the mechanical resonance (vibration) of a quartz crystal. An AC voltage applied at the resonance frequency of the quartz crystal causes it to continuously vibrate by the piezoelectric effect. The quartz oscillation frequency can be tuned from a few kHz to hundreds of MHz, depending on the size and shape of the quartz crystal and how it is cut. Due to the regularity of the vibration frequency, the quartz crystals can be used for time-keeping and are much more accurate than pendulums or other mechanical oscillators. They have a typical accuracy on the order of 1 second per month (i.e., they will gain or lose on the order of 1 second per month).

© The Author(s), under exclusive license to Springer Nature Switzerland AG 2022
R. LaPierre, *Getting Started in Quantum Optics*, Undergraduate Texts in Physics,
https://doi.org/10.1007/978-3-031-12432-7_20

20.2   Resonance Frequency in the Cs Atom

Many modern applications require clocks that are much more accurate than the quartz clock, which spurred the development of the atomic clock. The atomic clock is based on the electron transition between two energy levels in the ^{133}Cs atom. These two levels, labelled E_0 and E_1 in Fig. 20.1, are associated with the hyperfine splitting of the valence electron energy of the ^{133}Cs atom with principal quantum number $n = 6$ and orbital angular momentum quantum number $l = 0$. The hyperfine splitting is due to the Zeeman effect because of the magnetic field from the nuclear magnetic moment of the atom. The valence electron can occupy the lower energy level with spin down $(-\hbar/2)$ and the upper energy level with spin up $(+\hbar/2)$.

A Cs atom with an electron prepared in the E_0 state will absorb a photon when it is subjected to an incident electromagnetic wave tuned to the resonance frequency, f_0, of the two levels according to

$$f_0 = \frac{E_1 - E_0}{h} \tag{20.1}$$

Continued application of the incident field causes stimulated emission of a photon as the electron transitions back to E_0 from E_1. The incident field causes alternating absorption and emission of a photon.

According to the international system of units (Système International d'unités, informally known as the metric system), the second is defined by taking "the fixed numerical value of the cesium frequency, f_0, the unperturbed ground-state hyperfine transition frequency of the cesium-133 atom, to be 9,192,631,770 when expressed in the unit Hz, which is equal to s^{-1}". The frequency f_0 is defined to be exactly 9,192,631,770 Hz (i.e., in the microwave range) to closely match older definitions of the second.

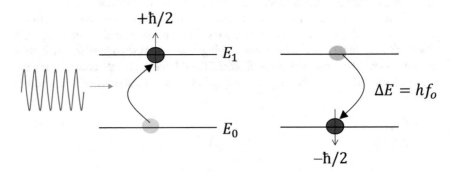

Fig. 20.1 Resonant absorption of a photon (depicted as the field in red) causes an electron transition from E_0 (with spin down) to E_1 (with spin up) as shown on the left. Continued application of the field causes stimulated emission of a photon and electron transition from E_1 to E_0 as shown on the right

20.3   Stern–Gerlach Apparatus

The atomic clock relies on an accurate measurement of the Cs transition frequency, f_0. This relies on the quantum state preparation of Cs atoms (i.e., preparing Cs atoms in the ground state, E_0). One way of performing the state preparation is the Stern–Gerlach (SG) apparatus, shown in Fig. 20.2. The SG experiment was first performed in 1922 by Otto Stern (Nobel Prize in Physics in 1943) and Walther Gerlach. In the SG experiment, a beam of atoms is produced by an oven and collimated by a slit (Ag atoms were used in the original SG experiment, while Cs atoms are used today for the atomic clock). The beam of atoms is passed through an inhomogeneous magnetic field produced by the shaped poles of a permanent magnet. This field will interact with the magnetic dipole moment of the atom, if any, and deflect it. This is easy to understand by consideration of the classical expression for the potential energy, U, of a magnetic dipole moment oriented along a magnetic field, B. If the magnetic field is predominantly along the z direction, then

$$U = -\mu_z B_z \tag{20.2}$$

The force on the magnetic moment is

$$F_z = -\frac{\partial U}{\partial z} = \mu_z \frac{\partial B_z}{\partial z} \tag{20.3}$$

The dipole moment will be deflected in the direction that decreases its potential energy.

According to Eq. (20.2), a magnetic moment parallel to the magnetic field can decrease its energy by moving to regions of higher magnetic field, while a magnetic moment pointing opposite the direction of the magnetic field can decrease its energy by moving to regions of lower magnetic field. Thus, the Cs atoms passing through the SG apparatus will be split into two beams, one having atoms with spin-up valence electron and the other having atoms with spin-down valence electron. By

Fig. 20.2 Schematic of the Stern–Gerlach apparatus

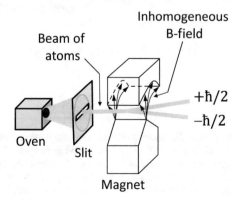

blocking one of the two beams and allowing the other beam to pass, the SG apparatus can be used as a state selection machine; that is, the SG apparatus is a filter for electron spin. An alternative method for state selection, which will not be discussed here, is optical pumping (optical pumping is described briefly in Chap. 21).

Exercise 20.1 Explain the physics responsible for deflection of Cs atoms into two separate beams in the Stern–Gerlach apparatus, as shown in Fig. 20.2.

20.4   Thermal Atomic Clock

The thermal atomic clock (Fig. 20.3) is based on thermally generated beams of Cs atoms. Cs is chosen because it has a low melting point making it easy to form a vapor, it has only one stable isotope (the hyperfine splitting will be identical for all atoms), it has a low ionization energy (it is easily ionized for detection), and it has a large hyperfine splitting due to a large nuclear spin.

The Cs beam passes through a SG apparatus allowing only Cs atoms with spin down (energy E_0) to pass. Then a microwave field is applied to flip the spin by resonant absorption to the level E_1. Another spin filter (SG apparatus) allows only atoms with spin up to pass to a detector. The Cs atoms are then detected, for example, by ionization followed by an ion detector or by state-dependent fluorescence emission. The detector signal is maximum when the applied microwave frequency exactly matches (is resonant with) the Cs transition frequency. The detector signal is used in a feedback loop to lock the microwave oscillator to the

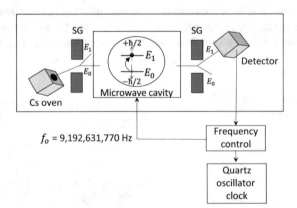

Fig. 20.3 The thermal atomic clock. An oven and collimating slit produce a beam of Cs atoms. A Stern–Gerlach (SG) apparatus is used to prepare all atoms in the spin-down state with energy E_0 before entering a microwave cavity. The microwave field of the cavity is resonantly absorbed by the Cs atoms causing a spin flip to the energy E_1. A second SG filter only allows atoms in the E_1 (spin up) state to pass to a detector. The detector signal is used as feedback for the frequency control

Cs transition frequency, which in turn can be used to provide a very stable AC voltage for a quartz crystal oscillator clock. Note that this description of the atomic clock is somewhat simplified—in practice, the atomic transition (spin flip) is achieved using a Ramsey interferometer. The interested reader may obtain further information from Refs. [1, 2].

> **Exercise 20.2** Investigate and explain the working principle of the Ramsey interferometer.

20.5   Improvements to the Atomic Clock

The accuracy of the atomic clock (or any clock) can be quantified by the Q factor, defined as follows:

$$Q = \frac{f_0}{\Delta f} \qquad (20.4)$$

where Δf is the spread or uncertainty in the transition frequency. A perfect clock would have a perfectly defined frequency f_0 with zero uncertainty ($\Delta f = 0$), giving an ideally infinite Q. Since all atoms of Cs are identical, the atomic oscillator essentially forms an almost perfect oscillator with a high Q. The Q for a thermal atomic clock is $\sim 10^{10}$, which is much better than quartz oscillators that have $Q \sim 10^4 - 10^6$. Δf (and hence Q) is ultimately limited by the natural linewidth of the atomic transition. In practice, another contribution to Δf in the thermal atomic clock is the Doppler effect that shifts the atomic transition frequency due to Doppler broadening because of the different speeds of the atoms exiting the oven. In addition, the Cs atoms pass quickly through the microwave cavity, limiting the interaction time. According to the frequency-time uncertainty relation, the limited interaction time leads to a spread in frequency.

Better clocks are made by trapping and cooling the atoms to reduce the Doppler broadening and increase the interaction time. Atom trapping and cooling are treated in the next chapter. Cs fountain clocks are based on the magneto-optical trapping of Cs atoms and can realize an uncertainty of 1 s in several hundred million years.

> **Exercise 20.3** Investigate and explain the working principle of the Cs fountain clock.

The next evolution of the atomic clock is to use atomic transitions in the optical rather than the microwave range, using atoms such as ^{199}Hg, ^{27}Al, ^{40}Ca, ^{174}Yb and ^{87}Sr. The microwave field would be replaced by an optical laser to induce the atomic transition. Optical frequencies are $\sim 10^5$ times higher than microwave frequencies

(THz versus GHz). This will further increase the Q factor and allow the measurement of smaller time intervals, making huge reductions in uncertainty possible. The measured frequencies of today's optical clocks are estimated to gain or lose no more than 1 s over the age of the universe!

20.6   Applications of the Atomic Clock

Coordinated Universal Time (UTC) serves as the official time reference for most of the world. UTC is computed from a weighted average of a network of nearly 450 atomic clocks located around the world. Computers, phones, and other devices have internal clocks (quartz oscillators) that need periodic correction. The clocks are synchronized to UTC using network time protocol.

Another application of the atomic clock is the Global Positioning System (GPS). GPS is a global navigation satellite system based on triangulation and transit time measurements. The satellites carry atomic clocks for precise measurements of the transit times.

The frequency of atomic clocks can be altered slightly by gravity (due to general relativity), magnetic fields (due to the Zeeman effect), electric fields (due to the Stark effect), and other phenomena. This sensitivity enables atomic clocks as ultra-precise measurement tools used, for example, in precision tests of special and general relativity, probing the merger of quantum mechanics and relativity, or tools in metrology [3, 4].

Exercise 20.4 Due to the accuracy of atomic clocks, it is desirable to express other measurement units (e.g., the meter) in terms of time (the second). Investigate the Système International d'unités (SI, informally known as the metric system), and establish which other units depend on the definition of the second.

References

1. E.F. Arias and G. Petit, *The hyperfine transition for the definition of the second*, Ann. Phys. (Berlin) 531 (2019) 1900068.
2. E.O. Gobel and U. Siegner, *The new international system of units (SI): Quantum metrology and quantum standards* (Wiley-VCH, 2019).
3. T. Bothwell et al., *Resolving the gravitational redshift across a millimetre-scale atomic sample*, Nature 602 (2022) 420.
4. B. Stray et al., *Quantum sensing for gravity cartography*, Nature 602 (2022) 590.

Chapter 21
Atom Cooling and Trapping

In recent decades, it has become possible to trap individual atoms and ions. These methods enable the interaction of light with a cloud of atoms or even with individual atoms. There are three general methods that are widely used to trap atoms or ions: radiation pressure, magnetic dipole forces, and electric dipole forces. In this chapter, the major trapping and cooling techniques are reviewed with an emphasis on the basic physical principles.

21.1 Paul Trap

Ions can be trapped in a "Paul trap", invented in the 1950s by Wolfgang Paul (Nobel Prize in Physics in 1945) [1]. According to Gauss's law, it is impossible to trap a single charge along all three directions in free space by static electric fields alone (Earnshaw's theorem), since there can be no net inward force ($\nabla \cdot E = 0$) to constrain the motion of the ions. There will be at least one direction where ions can escape. We can, however, use time varying electromagnetic fields to trap charge.

Before atomic trapping, the ions are first prepared as an atomic vapor, produced simply by evaporating the material in a vacuum chamber. These atoms can be ionized by an electron beam, laser beam, or high electric field to strip an electron off the atom. Be^+, Mg^+, Ca^+, Sr^+, Ba^+, Zn^+, Cd^+, Hg^+ and Yb^+ are commonly used ions, which have single valence electrons after ionization.

Figures 21.1 and 21.2 illustrate a Paul trap. An alternating potential is applied to electrodes, resulting in a saddle-shaped potential that rotates at the AC frequency (typically radio frequencies). The alternating forces create a trap for ions. Coulomb repulsion of the ions distributes them in a linear chain (1D crystal) along the trap (perpendicular to the page in Fig. 21.2). In 2016, more than 200 Be ions were trapped [2]. In 2018, a quantum register of 20 trapped ions were entangled [3]. Microfabricated 2D ion trap arrays are under development where RF and DC

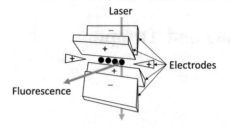

Fig. 21.1 Schematic of a Paul trap. The ions are shown as black dots. The potential on the four large electrodes alternates to create a "rotating saddle potential" that traps the ions. A laser is used to excite Rabi oscillations. Fluorescence from the ions can be read by a CCD camera or photodetector to determine the state of the ions

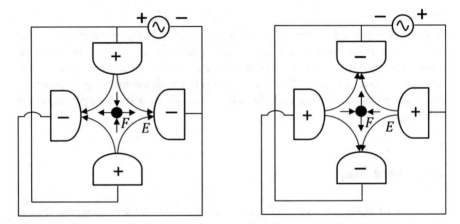

Fig. 21.2 A Paul trap illustrating the alternating potential at two different times (separated by a half-period of the AC potential), and the resulting electric field lines (E) and forces (F) on a positively charged ion. The resulting saddle potential rotates, resulting in an ion trap

fields can be applied to planar electrodes to move the ions around on the chip (ion shuttling) like electrons in a CCD camera [4].

21.2   Laser Cooling

Although fields are used to trap the ions, an alternative method is needed to reduce their energy (i.e., to cool them), so that the thermal energy does not cause unwanted atomic transitions. For this purpose, the method of laser or Doppler cooling is used, which works for both ions and neutral atoms with two electronic energy levels. As its name suggests, Doppler cooling involves the Doppler effect.

A laser of frequency ω_l is red-detuned from an electronic energy transition of the atom, meaning its frequency is slightly lower than the transition frequency ω_a of the

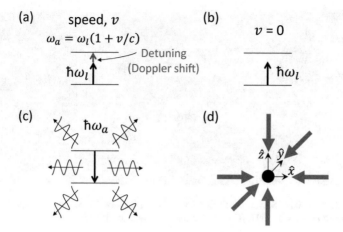

Fig. 21.3 Principle behind Doppler cooling. (**a**) Atomic motion toward the laser beam results in a Doppler shift of the photon frequency and resonant absorption. (**b**) Other atomic motion, such as a stationary atom, results in no photon absorption. (**c**) Each spontaneous emission occurs in a random direction, with momentum averaging to zero. (**d**) Three pairs of orthogonal counterpropagating laser beams will cool the atom along all three directions $(\hat{x}, \hat{y}, \hat{z})$

atom. An atom approaching the laser source with speed v will observe in its frame of reference a higher photon frequency due to the Doppler effect:

$$\omega_{\mathrm{a}} = \omega_{\mathrm{l}}\left(1 + \frac{v}{c}\right) \tag{21.1}$$

where c is the speed of light; that is, the Doppler shift is $\omega_{\mathrm{l}}\frac{v}{c} = kv$. The Doppler-shifted frequency becomes resonant with the electronic transition of the approaching atom, meaning the atom will absorb the photon (Fig. 21.3a). The momentum of the photon ($p = h/\lambda$) is transferred to the atom, giving it a "kick" in a direction that is opposite to the atomic motion. Atomic motion away from the laser beam, or a stationary atom, will not absorb the photon because $\omega_{\mathrm{a}} \neq \omega_{\mathrm{l}}$ (Fig. 21.3b). After photon absorption, the atom will move back into its ground state by spontaneous emission of the absorbed photon. The spontaneous emission occurs in a random direction (Fig. 21.3c). Hence, after many repeated absorption and emission events, the emission recoil momentum averages to zero, but the absorption recoil momentum does not. The emitted photon has energy $\hbar\omega_{\mathrm{a}}$, while the absorbed photon has energy $\hbar\omega_{\mathrm{l}} < \hbar\omega_{\mathrm{a}}$. As a result, the atom momentum and kinetic energy are reduced, cooling the atom to lower temperatures (the velocity distribution gets compressed). By using three pairs of laser beams oriented along orthogonal directions (Fig. 21.3d), the laser cooling can be applied along all three dimensions. This method of laser cooling is also called "optical molasses", since the lasers act as a viscous force that slows the atoms (it can be shown that $F \propto -v$, like a classical damping force, because the Doppler shift is proportional to velocity).

Fig. 21.4 Left to right: Steven Chu [5], Claude Cohen-Tannoudji [6], and William Phillips [7]; Nobel Prize in Physics in 1997. (Credit: Wikimedia Commons [5–7])

Note that Doppler cooling does not provide any restoring force; that is, it does not trap atoms but only cools them. Other methods are needed (e.g., a Paul trap) in combination with laser cooling to both trap and cool atoms. The limiting temperature (called the Doppler limit, T_D) is determined by the random walk of the atom caused by spontaneous emission. The Doppler limit is given by $T_D = \hbar\gamma/2k_B$ where γ is the rate of spontaneous emission (γ^{-1} is the natural lifetime of the excited state). The factor of 1/2 arises from the equal occupancy of the ground and excited states at high light intensities where stimulated emission balances absorption. For typical values of the natural linewidth, the temperature T_D is on the order of 100 μK, which is well below the temperature that can be achieved by cryogenic cooling methods. Steven Chu, Claude Cohen-Tannoudji, and William Phillips (Fig. 21.4) were awarded the 1997 Nobel Prize in Physics for their work on laser cooling and atom trapping.

21.3 Magneto-optical Trap

A Paul trap can trap ions, but this approach will not work for neutral atoms. Alternatively, we can use magneto-optical forces to trap and cool neutral atoms. Classically, the potential energy of a magnetic moment in a magnetic field is

$$U = -\boldsymbol{\mu} \cdot \boldsymbol{B} \qquad (21.2)$$

This results in a shift of the energy levels of the atom, known as the Zeeman effect, due to the interaction of the magnetic dipole moment of the atom with the applied magnetic field. Quantum mechanically, Eq. (21.2) is modified according to

$$U = -mg\mu_B B_z \qquad (21.3)$$

where m is the quantum number for the z-component of the total (orbital + spin) angular momentum, g is the Landé g-factor (a correction due to quantum electrodynamics; see Chap. 4), μ_B is the Bohr magneton, and B_z is the magnetic field along z. The magneto-optical trap (MOT), discussed below, uses the combined effects of an inhomogeneous magnetic field and an optical field to provide both cooling and trapping.

Exercise 21.1 Calculate the potential well depth, $U = \mu \Delta B$, for typical atomic magnetic moment values of the order of the Bohr magneton, $\mu \sim \mu_B$, and typical values of the variation in laboratory magnetic field, 0.01 T. Estimate the temperature at which atoms could be feasibly trapped.

In the MOT, an anti-Helmholtz coil is used to create an inhomogeneous magnetic field (a weak quadrupole magnetic field; Fig. 21.5a). The anti-Helmholtz coil consists of two solenoids of opposite current direction with axis aligned along the z-direction. These create a field-free region ($B = 0$ at the trap center, between the two coils) surrounded by regions of increasing magnetic field in all directions away from the trap center. An atom moving away from the zero-field region in the center of the trap toward, for example, the $+z$ or $-z$ direction between the two solenoids, will

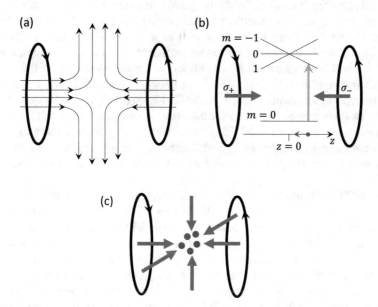

Fig. 21.5 (a) Anti-Helmholtz coil, showing a cross-section of the magnetic field lines. (b) Energy levels (ground state with $m = 0$, and Zeeman split excited state with $m = -1, 0, +1$); light polarization (σ_-, σ_+) of the laser beams (blue arrows); energy transition for absorption of σ_- photon (green arrow); and restoring force on an atom (red arrow). (c) Three pairs of laser beams provide 3D trapping

experience an increasing magnitude of the magnetic field ($\boldsymbol{B} = B_0 z \widehat{z}$) and a corresponding Zeeman shift of its energy levels.

Consider an atom with $m = 0$ for the ground state and three sublevels ($m = -1$, 0, $+1$) for the excited state. Thus, according to Eq. (21.3), there is no Zeeman splitting of the ground state, while the excited state is Zeeman split into three sublevels in the presence of a magnetic field. This energy splitting increases as the field increases, according to Eq. (21.3). The resulting energy levels are depicted in Fig. 21.5b.

A photon has spin angular momentum of $\hbar$, and projection of this angular momentum along the z axis (the propagation direction) gives $S_z = m\hbar$ where $m = -1$, 0, or $+1$, which is called left circularly polarized light (denoted σ_-), linearly polarized light (denoted Π), and right circularly polarized light (denoted σ_+), respectively. The absorption of light by an atom can only change m of the atom by -1, 0, or $+1$ due to conservation of angular momentum. σ_+ photons have $m = 1$ and can only increase m of an atom by one unit upon absorption of a photon. Conversely, σ_- photons have $m = -1$ and can only decrease m of the atom by one unit upon absorption of a photon. Finally, linearly polarized photons have $m = 0$ (they can be considered as a superposition of σ_+ and σ_- photons) and do not change m of the atom upon absorption.

In the MOT, two counterpropagating laser beams, one with σ_+ polarization and the other with σ_- polarization, are established along the z-axis for Doppler cooling of an atomic gas (Fig. 21.5b). σ_+ light propagates along the $+z$ direction, while σ_- light propagates along the $-z$ direction. An atom moving along $+z$ will have the Doppler-shifted σ_- light resonant with the $m = 0$ to $m = -1$ transition of the atom (Fig. 21.5b), while σ_+ light is far out of resonance. Conversely, an atom moving along $-z$ will be laser cooled by absorption of the σ_+ photon ($m = 0$ to $m = 1$ transition), but with negligible effect from the σ_- photon. Thus, the recoil of the atom from photon absorption is toward the field-free ($B = 0$) region located at $z = 0$, which provides a restoring force on the atom toward the trap center. The same principles apply along the $\widehat{x}$ and $\widehat{y}$ directions. By directing three counterpropagating laser beams along the three orthogonal directions, a centralized cloud of about 10^9 cold atoms can be trapped and cooled near the field-free region at the trap center (Fig. 21.5c).

The MOT is the most used cooling technique, allowing cooling of neutral atoms down to temperatures of $\sim 10 - 100 \ \mu K$. Among the new phenomenon enabled by a MOT is the Bose-Einstein condensate (BEC). The BEC is a group of bosonic atoms cooled to near absolute zero. A gas of atoms, such as Rb, is first laser cooled in a MOT followed by evaporative cooling. Evaporative cooling relies on the escape of high-energy atoms by lowering the trap potential, like cooling of a cup of coffee by blowing on it. Evaporative cooling can achieve nanoKelvin temperatures. When the atoms are cooled and trapped to a sufficient density (n), their de Broglie wavelengths (λ) become long enough to begin overlapping. When the condition $n\lambda^3 \sim 1$ is met, the atoms undergo a phase transition and enter a quantum degenerate state, which is called a BEC as predicted by Einstein in 1924. All atoms of the BEC can be described by a single wavefunction and act essentially as one single atom. Under these conditions, one has a coherent source of atoms enabling atomic interferometry, atomic lasers, and other coherent phenomenon, providing new tools for ultraprecise metrology. The first BEC was realized in 1995 [8] using Rb atoms and later Na atoms [9], which earned Eric A. Cornell, Wolfgang Ketterle, and Carl E. Wieman the Nobel Prize in Physics in 2001.

21.4 Sisyphus Cooling

Sisyphus cooling, also known as polarization gradient cooling, is a method of obtaining sub-Doppler cooling. Sisyphus cooling involves a combination of multilevel atoms, polarization gradients, light shifts, and optical pumping. Sub-Doppler cooling was discovered when temperature below the Doppler limit was realized in MOTs with atoms that have multiple levels in the ground state.

The method uses two counterpropagating laser beams with orthogonal linear polarizations, which may be described by

$$
\begin{aligned}
E &= E_0 \widehat{x} \cos{(\omega t - kz)} + E_0 \widehat{y} \cos{(\omega t + kz)} \\
&= E_0 (\widehat{x} + \widehat{y}) \cos{\omega t} \cos{kz} + E_0 (\widehat{x} - \widehat{y}) \sin{\omega t} \sin{kz}
\end{aligned}
\tag{21.4}
$$

Equation (21.4) describes a light field with a polarization gradient along $\widehat{z}$. For example, at $z = 0$, Eq. (21.4) gives $E = E_0 (\widehat{x} + \widehat{y}) \cos{(\omega t)}$, which describes linearly polarized light (denoted Π) at an angle of $\pi/4$ to the x-axis. Similarly, at $z = \lambda/4$, the light is linearly polarized but at an angle of $-\pi/4$ to the x-axis. At $z = \lambda/8$, the field becomes $E = \frac{E_0}{\sqrt{2}} (\widehat{x} + \widehat{y}) \cos{\omega t} + \frac{E_0}{\sqrt{2}} (\widehat{x} - \widehat{y}) \sin{\omega t}$. In the latter case, the sine and cosine terms give temporal dependences that are $\frac{\pi}{2}$ out of phase, which describes left circularly polarized light (denoted σ_-). Similarly, at $z = 3\lambda/8$, the light becomes right circularly polarized (denoted σ_+). Between the linear and circularly polarized light, the light is elliptically polarized. Thus, the light polarization oscillates between σ_- and σ_+ and back again over a distance of $\lambda/2$.

Consider an atom with possible states given by $|j, m\rangle$ where j is the total (spin + orbital) angular momentum and m is the total angular momentum quantized along the z-axis. For a given j, the allowed values of m are $m = 0, \pm 1, \pm 2, \ldots \pm j$, according to the rules of angular momentum in quantum mechanics. We suppose the atom has a ground state described by $j = \frac{1}{2}$ with two sublevels given by $m = \pm \frac{1}{2}$, and an excited state given by $j = \frac{3}{2}$ with 4 sublevels given by $m = -\frac{3}{2}, -\frac{1}{2}, \frac{1}{2}, \frac{3}{2}$.

Due to the conservation of angular momentum, the allowed transition with σ_+ light is from the $|\frac{1}{2}, \frac{1}{2}\rangle$ ground state to the $|\frac{3}{2}, \frac{3}{2}\rangle$ excited state, or from the $|\frac{1}{2}, -\frac{1}{2}\rangle$ ground state to the $|\frac{3}{2}, \frac{1}{2}\rangle$ excited state. The allowed transition with σ_- light is from $|\frac{1}{2}, \frac{1}{2}\rangle$ to $|\frac{3}{2}, -\frac{1}{2}\rangle$, or from $|\frac{1}{2}, -\frac{1}{2}\rangle$ to $|\frac{3}{2}, -\frac{3}{2}\rangle$. Finally, with linearly (Π) polarized light, only transitions from $|\frac{1}{2}, \frac{1}{2}\rangle$ to $|\frac{3}{2}, \frac{1}{2}\rangle$, or from $|\frac{1}{2}, -\frac{1}{2}\rangle$ to $|\frac{3}{2}, -\frac{1}{2}\rangle$ are allowed. The allowed transitions are illustrated in Fig. 21.6 along with the relative

Fig. 21.6 Allowed
transitions from ground to
excited state, showing the
photon polarizations
involved in the transitions.
The numbers are the
Clebsch–Gordan
coefficients, whose square
gives the transition
probabilities

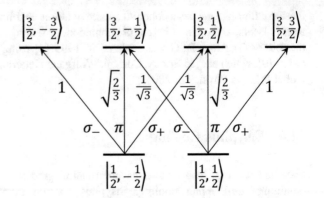

transition probabilities given by the square of the Clebsch–Gordan coefficients, according to the quantum mechanical theory of angular momentum.

The light shift (AC Stark effect) considered in Chap. 19 (Fig. 19.11) is dependent on the dipole coupling (g) between the ground and excited states. This coupling depends on the transition probabilities between the states (i.e., g is proportional to the square of the Clebsch–Gordan coefficients). Thus, the transition probabilities depend on the light polarization, as shown in Fig. 21.6. For example, examination of the Clebsch–Gordan coefficients in Fig. 21.6 shows that the transition from the $m = \frac{1}{2}$ state is three times more probable than from the $m = -\frac{1}{2}$ state for σ_+ polarization (i.e., the $m = \frac{1}{2}$ state is more strongly coupled to the light field). Thus, the light shift for σ_+ is three times greater when starting from $m = \frac{1}{2}$ compared to $m = -\frac{1}{2}$. Conversely, for σ_- polarization, the transition from the $m = -\frac{1}{2}$ state is three times more probable than from the $m = \frac{1}{2}$ state, and the corresponding light shift is three times greater. Finally, the transition probability and the light shift for linearly polarized light is equal from $m = \frac{1}{2}$ and $m = -\frac{1}{2}$, and equal to $\frac{2}{3}$ that of the maximum for circularly polarized light. In summary, the light shift of the ground state sublevels ($m = \pm\frac{1}{2}$) oscillates in space with the same periodicity as the polarization, as shown in Fig. 21.7.

An atom starting in the $m = \frac{1}{2}$ ground state sublevel and moving along z will lose kinetic energy (and gain potential energy) as it climbs the energy hill. Eventually, the atom encounters σ_- polarization, allowing a transition from the $\left|\frac{1}{2}, \frac{1}{2}\right\rangle$ ground state to the $\left|\frac{3}{2}, -\frac{1}{2}\right\rangle$ excited state (with $\Delta m = -1$). The atom will then spontaneously emit a photon and decay to the lower $m = -\frac{1}{2}$ ground state sublevel, losing energy in the process. This transition process is known as "optical pumping". The potential energy gained by the atom in climbing the hill is radiated away by the spontaneous emission, because the frequency of emission is greater than the frequency of absorption (see Fig. 21.7). The amount of energy lost is equal to the difference in light shift. As the atom continues moving along z, now in the $m = -\frac{1}{2}$ ground state sublevel, it again climbs the energy hill and next encounters σ_+ polarization, where it can again undergo optical pumping but now from the $\left|\frac{1}{2}, -\frac{1}{2}\right\rangle$ ground state to the $\left|\frac{3}{2}, \frac{1}{2}\right\rangle$ excited state ($\Delta m = +1$), and then back to the lower $\left|\frac{1}{2}, \frac{1}{2}\right\rangle$ ground state.

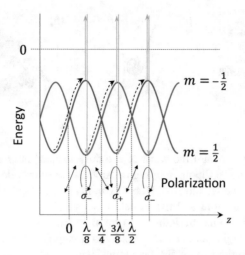

Fig. 21.7 Principle of Sisyphus cooling. Ground state energy levels with $m = \frac{1}{2}$ (red) and $m = -\frac{1}{2}$ (blue) are affected by the polarization-dependent light shift (the light shift is shown relative to zero energy). An atom starting at $z = 0$ in the $m = \frac{1}{2}$ ground state sublevel and moving along the z direction climbs uphill in energy (black dashed arrow). Optical pumping (green arrows) occurs upon absorption of a σ_- polarized photon. The process repeats in the $m = -\frac{1}{2}$ ground state sublevel with absorption of a σ_+ polarized photon

The process repeats along the z axis, resulting in a continuous loss of energy and sub-Doppler cooling to temperature on the order of μK. The name "Sisyphus cooling" derives from an analogy of the above process to a figure in Greek mythology who was doomed to forever roll a stone up a hill only to have it roll down again.

21.5   Dipole Trap, Optical Tweezers, and Optical Lattice

It is also possible to create a trapping potential using an induced electric dipole moment in an inhomogeneous electric field. Classically, the potential energy of an electric dipole moment in an electric field is given by

$$U = -\boldsymbol{p} \cdot \boldsymbol{E} \tag{21.5}$$

In an inhomogeneous field along z, the force is

$$F_z = -\frac{\partial U}{\partial z} = \frac{\partial (p_z E_z)}{\partial z} \tag{21.6}$$

The induced dipole moment is proportional to electric field, according to $p_z = \alpha E_z$ where α is the atomic polarizability. Thus, the force is proportional to the gradient of the light intensity, I (since $I \propto E^2$). This produces a force toward regions of higher

Fig. 21.8 (**a**) An atom trapped in the focus of a laser beam. (**b**) Atoms trapped in a 2D optical lattice. (Credit: Wikimedia Commons [12])

(a)   (b)

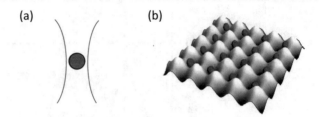

light intensity (e.g., toward the waist of a tightly focused laser beam; Fig. 21.8a). This approach is called dipole trapping or "optical tweezers", developed by Arthur Ashkin who was awarded the 2018 Nobel Prize in Physics. Optical tweezers are now an important tool to hold and manipulate microscopic objects such as biological molecules or living cells. In addition to trapping microscopic objects, the dipole interaction with a light field can also hold and manipulate atoms. The first optical trap was able to confine about 500 Na atoms [10].

In 1968, V.S. Letokhov proposed using the dipole force that arises from the light shift in a standing wave to confine atoms in microscopic dimensions [11]. Using a pattern of crossed laser beams, it is possible to set up a standing wave interference pattern with maxima and minima in the light intensity with a period on the order of the laser wavelength. The AC Stark effect or light shift varies with the light intensity, resulting in a gradient force on an atom given by $F_z = -\frac{\partial(\Delta E)}{\partial z}$ where ΔE is the spatially varying light shift. Atoms immersed in a laser field, which is red-detuned below atomic resonance (like in Fig. 19.11), have their ground state energy shifted down with increasing light intensity (negative light shift), and atoms are therefore attracted toward locations with maximum light intensity (toward the antinodes of the standing wave). The opposite occurs if the optical field is blue-detuned where atoms are attracted to the locations of minimum light intensity (nodes of the standing light wave). Using orthogonal laser beams, a 2D trap potential called an "optical lattice" can be created that enables dipole trapping of individual atoms at the nodes or antinodes of the light field in a 2D array like an egg carton (Fig. 21.8b). The trap potentials of an optical lattice are rather weak (a few mK), so it is necessary to first precool the atoms (e.g., to μK temperature in a MOT) and then superimpose the optical lattice to rearrange the atoms into the array.

The optical lattice is analogous to a conventional atomic crystal, but with a periodic potential of a few microns, about 10^3 times larger than the period of an atomic crystal. The de Broglie wavelength of the cooled atoms is on the order of the size of the trap, so the atomic motion in the trap must be described quantum mechanically. Atoms can tunnel between the potential wells of the optical lattice just like electrons in a crystal, and the motion of atoms through the periodic potential of an optical lattice can be described by a band structure derived from the Kronig–Penney model familiar from introductory quantum mechanics or solid-state physics [13]. Unlike an atomic crystal, however, we are free to vary the depth, period, or geometry of the potential wells by adjusting the light intensity, wavelength, or angles between the light beams. Thus, the array of atoms can be used to perform quantum simulations; that is, using one quantum system (the optical lattice) to simulate another one (like condensed matter systems such as superconductivity). One could also use optical lattices to control interactions between neighboring atoms and perform "microchemistry". Using light fields, it is possible to manipulate and steer atoms in a manner similar to mirrors, lenses, diffraction gratings, and other optical elements, creating a new domain of "atom optics". Such atom manipulation opens the possibility for atom-by-atom nanofabrication. It is even possible to trap and manipulate atoms into complex three-dimensional arrangements (Fig. 21.9) [14].

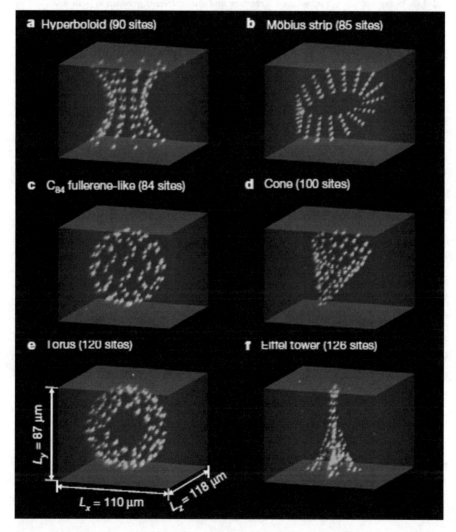

Fig. 21.9 Fluorescence from Rb-87 atoms arranged into 3D arrays. (Reprinted by permission from Springer Nature, Barredo et al. [14]. Copyright 2018)

Such arrangements could be used for quantum computing. A review of these and other aspects of atom trapping and cooling is provided in the classic reference by Metcalf and Straten [15].

References

1. W. Paul, *Electromagnetic traps for charged and neutral particles*, Rev. Mod. Phys. 62 (1990) 531.
2. J.G. Bohnet et al., *Quantum spin dynamics and entanglement generation with hundreds of trapped ions*, Science 352 (2016) 1297.
3. N. Friis et al., *Observation of entangled states of a fully controlled 20-qubit system*, Phys. Rev. X 8 (2018) 021012.
4. C.D. Bruzewicz et al., *Trapped-ion quantum computing: Progress and challenges*, Appl. Phys. Rev. 6 (2019) 021314.
5. Attribution: The Royal Society. This file is licensed under the Creative Commons Attribution-Share Alike 3.0 Unported license (https://creativecommons.org/licenses/by-sa/3.0/deed.en). File: Professor Steven Chu ForMemRS headshot.jpg. (2020, October 30). *Wikimedia Commons, the free media repository*. Retrieved 18:38, December 8, 2020 from https://commons.wikimedia.org/w/index.php?title=File:Professor_Steven_Chu_ForMemRS_headshot.jpg&oldid=507616885
6. Author: Amir Bernat. This file is licensed under the Creative Commons Attribution-Share Alike 4.0 International, 3.0 Unported, 2.5 Generic, 2.0 Generic and 1.0 Generic license (https://creativecommons.org/licenses/by-sa/4.0/). File:Claude Cohen-Tannoudji.JPG. (2020, August 28). *Wikimedia Commons, the free media repository*. Retrieved 18:06, December 8, 2020 from https://commons.wikimedia.org/w/index.php?title=File:Claude_Cohen-Tannoudji.JPG&oldid=444509221
7. Author: Markus Pössel. This file is licensed under the Creative Commons Attribution-Share Alike 3.0 Unported license (https://creativecommons.org/licenses/by-sa/3.0/deed.en). File: William D. Phillips.jpg. (2020, October 26). *Wikimedia Commons, the free media repository*. Retrieved 18:08, December 8, 2020 from https://commons.wikimedia.org/w/index.php?title=File:William_D._Phillips.jpg&oldid=502669574.
8. M.H. Anderson, J.R. Ensher, M.R. Matthews, C.E. Wieman and E.A. Cornell, *Observation of Bose-Einstein condensation in a dilute atomic vapor*, Science 269 (1995) 198.
9. K.B. Davis, M.-O. Mewes, M.R. Andrews, M.J. Van Druten, D.S. Durfee, D.M. Kurn and W. Ketterle, *Bose-Einstein condensation in a gas of sodium atoms*, Phys. Rev. Lett. 75 (1995) 3969.
10. S. Chu, J.E. Bjorkholm, A. Ashkin and A. Cable, *Experimental observation of optically trapped atoms*, Phys. Rev. Lett. 57 (1986) 314.
11. V.S. Lethokov, *Narrowing of the Doppler width in a standing light wave*, JETP Lett. 7 (1968) 272.
12. https://commons.wikimedia.org/wiki/File:Lattice_mott.JPG
13. Y. Castin and J. Dalibard, *Quantization of atomic motion in optical molasses*, Europhys. Lett. 14 (1991) 761.
14. D. Barredo et al., *Synthetic three-dimensional atomic structures assembled atom by atom*, Nature 561 (2018) 79.
15. H.J. Metcalf and P. van der Straten, *Laser cooling and trapping* (1999, Springer).

Further Reading

H.-A. Bachor and T.C. Ralph, *A guide to experiments in quantum optics* (2019, Wiley-VCH, 3rd ed.).

C.C. Gerry and P.L. Knight, *Introductory quantum optics* (2005, Cambridge University Press).

G. Grynberg, A. Aspect and C. Fabre, *Introduction to quantum optics* (2010, Cambridge University Press).

R.R. LaPierre, *Introduction to quantum computing* (2021, Springer).

P. Meystre, *Quantum optics* (2021, Springer).

H.J. Metcalf and P. van der Straten, *Laser cooling and trapping* (1999, Springer).

© The Editor(s) (if applicable) and The Author(s), under exclusive license to
Springer Nature Switzerland AG 2022
R. LaPierre, *Getting Started in Quantum Optics*, Undergraduate Texts in Physics,
https://doi.org/10.1007/978-3-031-12432-7

Appendix 1: Derivation of Lamb Shift

A heuristic derivation of the Lamb shift was put forward by Theodore A. Welton in 1948 [A1.1]. We assume our atom sits inside a box containing the electric field of the vacuum modes, like Fig. 3.4. These electric fields of the vacuum modes shake the electron. The average energy density (energy per unit volume) inside the box associated with each mode k of the electric field (E_k) is

$$\langle U_k \rangle = \frac{1}{2}\epsilon_0 \langle (E_k)^2 \rangle = \frac{1}{2}\epsilon_0 \left(\varepsilon^1\right)^2 = \frac{1}{2}\epsilon_0 \left(\sqrt{\frac{\hbar\omega}{2\epsilon_0 V}}\right)^2 = \frac{1}{4}\frac{\hbar\omega}{V} \tag{A1.1}$$

where ε^1 is the one-photon field amplitude of the mode k with frequency $\omega = ck$. The energy associated with the electric field is equal to the energy density multiplied by the volume of the box:

$$\langle U_k \rangle V = \langle U_k \rangle L^3 = \frac{1}{4}\hbar\omega = \frac{1}{4}\hbar ck \tag{A1.2}$$

Equation (A1.2) simply tells us that the ground state energy of each vacuum mode, $\frac{1}{2}\hbar\omega$, is divided equally between the electric and magnetic fields. Thus,

$$\langle (E_k)^2 \rangle = \frac{\hbar ck}{2\epsilon_0 V} \tag{A1.3}$$

The acceleration of the electron due to each electric field of mode k is

$$a_k = \frac{F}{m} = \frac{eE_k}{m} \tag{A1.4}$$

The oscillatory motion of the electron is given by

© The Editor(s) (if applicable) and The Author(s), under exclusive license to
Springer Nature Switzerland AG 2022
R. LaPierre, *Getting Started in Quantum Optics*, Undergraduate Texts in Physics,
https://doi.org/10.1007/978-3-031-12432-7

$$\delta r_k(t) = \delta r_k \cos(\omega t) \tag{A1.5}$$

The acceleration, a_k, is given by

$$a_k(t) = \frac{d^2(\delta r_k(t))}{dt^2} = -\omega^2 \delta r_k(t) = -k^2 c^2 \delta r_k(t) \tag{A1.6}$$

Thus, from Eqs. (A1.4) and (A1.6), the displacement amplitude is

$$\delta r_k = \frac{e}{mc^2} E_k \frac{1}{k^2} \tag{A1.7}$$

The mean square displacement from Eqs. (A1.7) and (A1.3) is

$$\langle (\delta r_k)^2 \rangle = \left(\frac{e}{mc^2}\right)^2 \langle (E_k)^2 \rangle \frac{1}{k^4} = \left(\frac{e}{mc^2}\right)^2 \left(\frac{\hbar c}{2\epsilon_0 V}\right) \frac{1}{k^3} \tag{A1.8}$$

The total mean square displacement $\langle (\delta r)^2 \rangle$ is a summation of $\langle (\delta r_k)^2 \rangle$ over all modes k:

$$\langle (\delta r)^2 \rangle = \sum_k \langle (\delta r_k)^2 \rangle \tag{A1.9}$$

$$= \left(\frac{e}{mc^2}\right)^2 \left(\frac{\hbar c}{2\epsilon_0 V}\right) \sum_k \frac{1}{k^3} \tag{A1.10}$$

Note that Eq. (A1.9) is valid because random variables add in quadrature. This is the familiar rule for root-mean-square (rms) addition.

We assume the fictitious box is very large, so the modes are closely spaced in wavevector (k is continuous) and we can replace the summation with an integral:

$$\langle (\delta r)^2 \rangle = \left(\frac{e}{mc^2}\right)^2 \left(\frac{\hbar c}{2\epsilon_0 V}\right) \int_0^\infty \frac{\rho(k)}{k^3} d^3 k \tag{A1.11}$$

$\rho(k)$ is the density of modes, that is, the number of modes between k and $k + dk$ in k-space.

To determine $\rho(k)$, note that the wavevectors for waves in the box are quantized by periodic boundary conditions according to

$$k_x = \frac{2\pi}{\lambda} = \frac{2\pi}{L/n_x} = \frac{2\pi}{L} n_x \tag{A1.12}$$

Similarly,

$$k_y = \frac{2\pi}{L} n_y \qquad (A1.13)$$

$$k_z = \frac{2\pi}{L} n_z \qquad (A1.14)$$

Each mode takes up a volume in k-space of $\left(\frac{2\pi}{L}\right)^3$. The number of modes in the range k to $k + dk$ is

$$\rho(k)dk = \frac{2(4\pi k^2)dk}{\left(\frac{2\pi}{L}\right)^3} = \frac{L^3 k^2 dk}{\pi^2} \qquad (A1.15)$$

where the first factor of 2 arises from the two allowed polarizations.

Thus,

$$\langle(\delta r)^2\rangle = \left(\frac{e}{\pi mc^2}\right)^2 \left(\frac{\hbar c}{2\epsilon_0}\right) \int_0^\infty \frac{1}{k} dk \qquad (A1.16)$$

We see that the volume $V = L^3$ of the box cancels out as it should since the box was fictitious.

The integral in Eq. (A1.16) is divergent, so we need to choose suitable limits in the integration. The maximum wavelength is associated with the size of the atom, which is twice the Bohr radius, $2a_0 = \frac{2\epsilon_0 h^2}{\pi m e^2}$. This maximum wavelength gives the minimum wavevector:

$$k_{min} = \frac{2\pi}{2a_0} = \frac{\pi^2 m e^2}{\epsilon_0 h^2} \qquad (A1.17)$$

The minimum wavelength is taken to be the Compton wavelength, $\lambda_c = \frac{h}{mc}$. The Compton wavelength gives a fundamental limitation on measuring the position of a particle. This minimum wavelength gives the maximum wavevector:

$$k_{max} = \frac{2\pi}{\lambda_c} = \frac{2\pi mc}{h} \qquad (A1.18)$$

Thus, the integral becomes

$$\int_{k_{min}}^{k_{max}} \frac{1}{k} dk = \ln\left(\frac{k_{max}}{k_{min}}\right) = \ln\left(\frac{4\epsilon_0 \hbar c}{e^2}\right) = \ln\left(\frac{1}{\pi\alpha}\right) \qquad (A1.19)$$

where $\alpha = \frac{1}{4\pi\epsilon_0} \frac{e^2}{\hbar c} \approx \frac{1}{137}$ is the fine structure constant. Thus,

$$\langle(\delta r)^2\rangle = \left(\frac{e}{\pi mc^2}\right)^2\left(\frac{\hbar c}{2\epsilon_0}\right)\ln\left(\frac{1}{\pi\alpha}\right) \tag{A1.20}$$

Because of the electron displacement, the electron experiences a change in potential according to

$$\Delta V = V(r+\delta r) - V(r) = \delta r\nabla V + \frac{1}{2}(\delta r)^2\nabla^2 V + \dots \tag{A1.21}$$

where we have used a Taylor expansion. According to Eq. (A1.5), the average of δr is zero, leaving:

$$\Delta V = \frac{1}{2}(\delta r)^2\nabla^2 V \tag{A1.22}$$

Here V is the Coulomb potential of the hydrogen atom:

$$V = -\frac{1}{4\pi\epsilon_0}\frac{e^2}{r} \tag{A1.23}$$

The Laplacian, $\nabla^2 V$, for the Coulomb potential is

$$\nabla^2 V = \frac{e^2}{\epsilon_0}\delta(r) \tag{A1.24}$$

where $\delta(r)$ is the Dirac delta function. The average becomes

$$\langle\nabla^2 V\rangle = \langle\psi|\nabla^2 V|\psi\rangle \tag{A1.25}$$

where ψ is the hydrogen wavefunction. Using Eq. (A1.24), we get

$$\langle\nabla^2 V\rangle = \frac{e^2}{\epsilon_0}|\psi(0)|^2 \tag{A1.26}$$

Equation (A1.26) is zero at the origin for the p orbital and nonzero only for the s orbital. For the $2s$ orbital, the wavefunction at the origin is

$$\psi_{2s}(0) = \frac{1}{(8\pi a_0^3)^{1/2}} \tag{A1.27}$$

Thus,

$$\langle \Delta V \rangle = \frac{1}{2} \langle (\delta r)^2 \rangle \langle \nabla^2 V \rangle \tag{A1.28}$$

$$= \frac{1}{2\pi} \alpha^5 mc^2 \ln \left(\frac{1}{\pi \alpha} \right) \tag{A1.29}$$

This shift is about 1540 MHz, close to the observed shift of 1058 MHz. A more exact calculation using the full theory of quantum electrodynamics provides a measurement of the fine-structure constant α to better than one part in a million.

References

A1.1. T.A. Welton, *Some observable effects of the quantum-mechanical fluctuations of the electromagnetic field*, Phys. Rev. 74 (1948) 1157.

Appendix 2: Derivation of Casimir Formula

Here, we present a simple heuristic argument for the $1/d^4$ dependence of the Casimir force. The zero-point energy between the plates is

$$E = \frac{1}{2}\hbar\sum_n \omega_n \tag{A2.1}$$

$$= \frac{1}{2}\hbar c\sum_n k_n \tag{A2.2}$$

where c is the speed of light.

The wavevectors perpendicular to the plates, describing standing waves, are quantized according to

$$k_n = \frac{n\pi}{d} \tag{A2.3}$$

where n is an integer and d is the plate separation.

The density of states is $\rho(k)dk \propto (Ad)k^2 dk$ where A is the plate area, and Ad is the mode volume between the plates. Thus, we can replace the summation in Eq. (A2.2) by an integration, using the density of states:

$$E \propto Ad\hbar c\int_{k_{min}}^{k_{max}} k^3 dk \tag{A2.4}$$

According to Eq. (A2.3), the minimum wavevector is $k_{min} = \frac{\pi}{d}$. The maximum wavevector corresponds to the maximum allowed frequency, which is taken to be the plasma frequency of the metal where the plates become transparent. Thus, evaluating the integral gives

R. LaPierre, *Getting Started in Quantum Optics*, Undergraduate Texts in Physics, https://doi.org/10.1007/978-3-031-12432-7

$$E \propto Adhck^4 \Big|_{k_{min}}^{k_{max}}$$

(A2.5)

$$E \propto Adhc \left(k_{max}^4 - \left(\frac{\pi}{d}\right)^4 \right)$$

(A2.6)

$$E \propto -Ahc\frac{1}{d^3}$$

(A2.7)

The force per unit area F on the plates is

$$F = -\frac{1}{A}\frac{\partial E}{\partial d} \propto -\frac{\hbar c}{d^4}$$

(A2.8)

Equation (A2.8) gives the correct $1/d^4$ dependence of the Casimir force. The derivation is not quite correct because only modes with wavevectors perpendicular to the plates are quantized, and one must use periodic boundary conditions for modes with wavevectors parallel to the plates. A more detailed calculation [A2.1, A2.2] gives the correct numerical factor:

$$F = -\frac{\pi^2 \hbar c}{240}\frac{1}{d^4}$$

(A2.9)

References

A2.1. S. Scheel and S.Y. Buhmann, *Macroscopic quantum electrodynamics — concepts and applications*, Acta Phys. Slov. 58 (2008) 700.
A2.2. S.K. Lamoreaux, *The Casimir force: background, experiments, and applications*, Rep. Prog. Phys. 68 (2004) 201.

Appendix 3: Derivation of Normalization Constant in Single Photon Wavepacket

We start with the Lorentzian distribution of mode probabilities:

$$|c_l|^2 = \frac{K^2}{(\omega_l - \omega_0)^2 + \gamma^2/4} \tag{A3.1}$$

The normalization condition becomes

$$\sum_{l=0}^{\infty} |c_l|^2 = \sum_{l=0}^{\infty} \frac{K^2}{(\omega_l - \omega_0)^2 + \gamma^2/4} = 1 \tag{A3.2}$$

Let us replace the discrete summation with an integration:

$$\sum_{l=0}^{\infty} \frac{K^2}{(\omega_l - \omega_0)^2 + \gamma^2/4} = \int_0^{\infty} \frac{K^2}{(\omega_l - \omega_0)^2 + \gamma^2/4} \rho(\omega_l) d\omega_l \tag{A3.3}$$

where $\rho(\omega_l)$ is the density of modes, and $\rho(\omega_l)d\omega_l$ is the number of modes in the frequency interval from ω_l to $\omega_l + d\omega_l$.

Suppose we have a wavepacket of length L. The allowed wavevectors of the standing modes become discrete:

$$k_l = \frac{2\pi}{\lambda} = \frac{2\pi}{2L/n_l} = \frac{\pi}{L} n_l \tag{A3.4}$$

$$\omega_l = k_l c = \frac{\pi c}{L} n_l \tag{A3.5}$$

© The Editor(s) (if applicable) and The Author(s), under exclusive license to
Springer Nature Switzerland AG 2022
R. LaPierre, *Getting Started in Quantum Optics*, Undergraduate Texts in Physics,
https://doi.org/10.1007/978-3-031-12432-7

Rearranging,

$$\rho(\omega_l)d\omega = \frac{dn_l}{d\omega_l}d\omega = \frac{L}{\pi c}d\omega \qquad (A3.6)$$

Thus,

$$\int_0^\infty \frac{K^2}{(\omega_l - \omega_0)^2 + \gamma^2/4}\rho(\omega_l)d\omega_l = \frac{LK^2}{\pi c}\int_0^\infty \frac{1}{(\omega_l - \omega_0)^2 + \gamma^2/4}d\omega_l \qquad (A3.7)$$

$$= \frac{4LK^2}{\pi c\gamma^2}\int_0^\infty \frac{1}{4(\omega_l - \omega_0)^2/\gamma^2 + 1}d\omega_l \qquad (A3.8)$$

Let $u = 2(\omega_l - \omega_0)/\gamma$, $du = (2/\gamma)\, d\omega_l$, giving

$$\frac{4LK^2}{\pi c\gamma^2}\int_0^\infty \frac{1}{4(\omega_l - \omega_0)^2/\gamma^2 + 1}d\omega_l = \frac{2LK^2}{\pi c\gamma}\int_0^\infty \frac{1}{u^2 + 1}du \qquad (A3.9)$$

Using $\int_0^\infty \frac{1}{u^2+1}du = \tan^{-1}(u)\big|_0^\infty = \pi/2$ gives

$$\frac{2LK^2}{\pi c\gamma}\int_0^\infty \frac{1}{u^2 + 1}du = \frac{LK^2}{c\gamma} \qquad (A3.10)$$

Now we apply the normalization condition:

$$\frac{LK^2}{c\gamma} = 1 \qquad (A3.11)$$

which gives

$$K = \sqrt{\frac{c\gamma}{L}} \qquad (A3.12)$$

Appendix 4: Derivation of Planck's Distribution Law

Starting with Eq. (12.23):

$$P_n = e^{-n\hbar\omega/kT} \left(1 - e^{-\hbar\omega/kT}\right) \tag{A4.1}$$

The energy is given by

$$\langle E \rangle = \sum_{n=0}^{\infty} E_n P_n \tag{A4.2}$$

Ignoring the zero-point energy, we get

$$\langle E \rangle = \sum_{n=0}^{\infty} n\hbar\omega e^{-n\hbar\omega/kT} \left(1 - e^{-\hbar\omega/kT}\right) \tag{A4.3}$$

$$= \hbar\omega \left(1 - e^{-\hbar\omega/kT}\right) \sum_{n=0}^{\infty} n\, e^{-n\hbar\omega/kT} \tag{A4.4}$$

Let $a = e^{-\hbar\omega/kT}$, giving

$$\langle E \rangle = \hbar\omega \left(1 - a\right) \sum_{n=0}^{\infty} na^n \tag{A4.5}$$

Next, we use the geometric series $\sum_{n=1}^{\infty} na^n = \frac{a}{(1-a)^2}$, giving

$$\langle E \rangle = \hbar\omega \, \frac{a}{1 - a} \tag{A4.6}$$

© The Editor(s) (if applicable) and The Author(s), under exclusive license to
Springer Nature Switzerland AG 2022
R. LaPierre, *Getting Started in Quantum Optics*, Undergraduate Texts in Physics,
https://doi.org/10.1007/978-3-031-12432-7

$$= \hbar\omega \frac{e^{-\hbar\omega/kT}}{1 - e^{-\hbar\omega/kT}} \tag{A4.7}$$

$$= \frac{\hbar\omega}{e^{\hbar\omega/kT} - 1} \tag{A4.8}$$

Next, we need to find the density of modes; that is, how many modes exist with frequency between ω and $\omega + d\omega$. Expressing Eq. (A1.15) in terms of frequency ($k = \frac{\omega}{c}$), we get

$$\rho(\omega)d\omega = \frac{L^3}{\pi^2 c^3} \omega^2 d\omega \tag{A4.9}$$

The number of states per unit volume between frequency ω and $\omega + d\omega$ is

$$\frac{1}{L^3}\rho(\omega)d\omega = \frac{1}{\pi^2 c^3} \omega^2 d\omega \tag{A4.10}$$

Thus, the energy density between ω and $\omega + d\omega$ is

$$U(\omega)d\omega = \frac{1}{\pi^2 c^3} \omega^2 \frac{\hbar\omega}{e^{\hbar\omega/kT} - 1} d\omega \tag{A4.11}$$

$$U(\omega)d\omega = \frac{h}{2\pi^3 c^3} \frac{\omega^3}{e^{\hbar\omega/kT} - 1} d\omega \tag{A4.12}$$

Equation (A4.12) gives the Planck radiation law as

$$U(\omega) = \frac{h}{2\pi^3 c^3} \frac{\omega^3}{e^{\hbar\omega/kT} - 1} \tag{A4.13}$$

$U(\omega)$ is called the spectral energy density with units of Jm^{-3}/s^{-1}, that is, energy per unit volume per unit frequency. Thus, if we integrate Eq. (A4.13) with respect to frequency, ω, we get the energy density (energy per unit volume in units of Jm^{-3}). Similarly, the spectral energy density could be expressed in terms of wavelength:

$$U(\lambda) = \frac{8\pi hc}{\lambda^5} \frac{1}{e^{hc/\lambda kT} - 1} \tag{A4.14}$$

Classically, we expect kT of energy per mode of the radiation, giving

$$U(\omega) = \frac{1}{V}\rho(\omega)kT = \frac{\omega^2}{\pi^2 c^3} kT \tag{A4.15}$$

which is known as the Rayleigh–Jeans result. This classical result diverges with increasing ω, leading to the so-called uv catastrophe. Plank's result solves this problem by quantizing the radiation, leading to perfect agreement with experiment.

Index

Printed in the United States
by Baker & Taylor Publisher Services